세계유용곤충대도감시리즈 Ⅲ
# 세계의 꽃무지 대도감

저자 **손민우**

지은이

손민우 / TEL. 063)580-4815
전라북도 부안군 부안읍 동중리 222-1

펴낸이

(주)커뮤니케이션 열림 / TEL. 031)955-0123  FAX. 031)955-0119
경기도 파주시 교하읍 문발리 파주출판도시 514-7
www.comopen.co.kr

2009년 4월 10일 인쇄
2009년 4월 18일 발간

Author

Son Minwoo / TEL. +82-63-580-4815
222-1, Dongjung-ri, Buan-eup, Buan-gun, Jeollabuk-do,
SEOUL 579-700 KOREA.

Published by

Communication Yeollim Co., Ltd. / TEL. +82-31-955-0123  FAX. +82-31-955-0119
514-7, Paju Book City, Munbal-ri, Gyoha-eup, Paju-si,
Gyeonggi-do, SEOUL 413-756 KOREA.

First printed in 2009
First published in 2009

ISBN 978-89-959228-8-0-96490
ISBN 978-89-959228-4-2-96490(세트)
정가 38,000원

• 이 책은 저작권법에 의해 보호를 받는 저작물이므로 글과 사진의 무단전재와 복제를 금합니다.
All rights reserved. No part of this book may be used or reproduced in any manner whatsoever without written permission.

표지소재 협찬 : (주)윙윙아이디 (02-2264-1347)
소재명 : 윙윙 - 파피루스 레인 #04

세계유용곤충대도감시리즈 Ⅲ
Photograph Book Series of the World Insects Vol. Ⅲ

# 세계의 꽃무지 대도감

The Cetoniid Beetles of the World (Flower Beetles)

by Son Minwoo

2009

(주)커뮤니케이션 열림

Communication Yeollim Co., Ltd.

# 머리말

세계유용곤충대도감시리즈 1편 "세계의 사슴벌레 대도감"의 출간에 이어 이번에 시리즈 2편 "세계의 장수풍뎅이 대도감"과 3편 "세계의 꽃무지 대도감"을 출간하게 되었다.

"이 도감에는 나의 조국, 대한민국의 동해와 독도가 있다.
우리의 영토 우리의 영해 즉, 우리의 자주성을 지키기 위해서는 우리 스스로가 주체가 되어 세계 모든 이에게 알리고 인정받아야 한다."

20여 년 전 독도의 영유권 문제가 대두 했을 때 "우리의 것을 어떻게 지켜 나가야 할 것 인가?"를 고민하게 되었고, 개인적으로라도 우리만의 곤충도감을 제작해야 겠다고 결심했었다. 대다수 국역 외국곤충도감이 갖고 있는 왜곡된 번안으로부터 우리의 미래인 어린이들을 지키고 주체성을 확립하고자 시작한 일이 어느덧 20여 년의 세월이 흐른 것이다.

그 간의 인고를 생각하면 출간의 기쁨은 감출 수 없고, 가슴 속의 뜨거운 무언가가 목을 메이게 한다. 세 편의 도감을 출간하게 되었지만 아직도 시작에 불과하다는 초심으로 마지막까지 정진 해야겠다는 마음을 다잡는다.

지난 5년 간 세 편의 세계유용곤충대도감시리즈를 집필 하는 동안 많은 분들의 참여와 후원이 있었고 그리고 국가신활력사업의 사업비도 지원받게 되었다. 그 위에 그래픽과 인쇄, 출판 전문인들의 열정이 더해져 1편 도감이 출간되었고, 그해 2008년 "문화체육관광부의 우수학술도서"로 선정되는 기쁨까지도 누리기에 이르렀다.

지난 일들을 되돌아보면 지금까지 출간 된 도감에 600여종의 곤충사진을 올리기 위하여 무려 7,000여장의 사진을 촬영하였고, 곤충표본사진 한 장을 완성하기 위해 10여 장을 촬영하여, 일일이 핀트를 맞추고, 그 위에 리터칭과 교정을 반복 하고 또 반복 하였다. 밤낮없이 이어진 고된 작업이었지만 이 도감이 미래의 우리 어린이들에게 큰 자부심으로 남을 수 있다고 생각하면서 한 장 한 장의 원고에 심혈을 기울였다.

또한, 독자적으로 개발한 사진촬영법과 더불어 진보된 편집기술과 첨단의 인쇄기술로 어디에 내 놓아도 손색이 없을 정도의 입체감과 표면의 질감이 살아있는 세계최고의 고해상도 도감을 만들고자 하였으며, 이 책의 모든 내용은 세계화를 위해 국영 혼문판으로 편집하였고, 앞으로 각국의 언어로도 번역되어 전 세계 에서 출간되기를 희망한다.

이제 세계유용곤충 대도감시리즈 2편 "세계의 장수풍뎅이 대도감"과 3편 "세계의 꽃무지 대도감"으로 중간 결실을 보았으나, 곤충탐사와 채집의 어려움으로 인하여, 더 많은 종의 사진을 올리지 못한 아쉬움을 향후 발전의 계기로 남긴다.

이 책이 만들어 지기까지 도움을 주신 분들을 모두 나열할 수는 없지만, 이 도감이 세상에 나오도록 기회를 주신 부안군수님과 부안군 관계자 여러분들께 감사드리며, 감수를 맡아 주신 서지은박사님과 오성환박사님, 자문위원이신 박윤점교수님과 오순환교수님, 김태완소장님, 그래픽 디자인과 출판을 맡아 주신 (주)커뮤니케이션 열림의 박철영 대표님과 스텝진, 수고가 가장 많았던 표본사진 합성 및 트리밍을 맡은 이수진님께 감사드린다. 끝으로 그 동안 꾸준히 지켜봐 주시고 힘이 되어 주셨던 어머님과 고인이 되신 아버님, 원고 집필을 핑계로 매일 늦게 귀가하는 아버지를 끊임없이 성원해준 두 아들 창현과 창건에게도 고마움을 전한다.

2009. 4.  저자 손민우

# Preface

We release "The Dynastid Beetles of the World(Rhinoceros Beetles)", Useful Insects of the World series 2, and "The Cetoniid Beetles of the World(Flower Beetles)", Useful Insects of the World series 3, after release of "The Lucanid Beetles of the World", Useful Insects of the World series 1.

"This book has the East Sea and Dokdo in Korea, my home nation, as the background.
We should be the subject for recognition of all the people in the world in order to protect our independent territory and closed sea."

When issue about possession of Dokdo came up 20 years ago, I was worried about "how should we keep our property", and determined that I should make an photograph book about beetle by myself. It has passed about 20 years since I started to make a book in order to let children, our future, read the good book, not the existing bug books whose translation is terrible, and make them establish their subjecthood.

I am very delightful about publication of this book because my endurance is not in vain. Although I have releases three books, I determine to keep my original intention and do my best to the end.

A lot of persons have given help and support in writing of 3 books of Useful Beetles of the World series for the past 5 years, and government has provided aids in the name of new energy project. The first book of the series released with graphic, printing, and passion of specialized publishers, and I was delighted that the book was chosen as the 2008 Superior Scientific Book Selected by Ministry of Culture, Sports and Tourism in the year.

I took about 7 thousand pictures for beetle pictures of about 600 species in the 3 books, and for one perfect beetle picture I took 10 partical pictures and adjusted the focus respectively, and also repeated retouching and correcting it. Although I had to work hard day and night, I put my heart and soul into every single paper with the thought that our future children will be proud of this book.

Moreover, I used unique method of taking a picture, advanced editing technique and high-tech printing technology for the best HD photograph book of the World that has cubic effect and live feel of material on the face. All the sentences in the book edited to be Korean and English. I hope that this book will be release all over the world with language of each nation.

Although I am kind of satisfied with release of "The Dynastid Beetles of the World(Rhinoceros Beetles)", Useful Insects of the World series 2, and "The Cetoniid Beetles of the World(Flower Beetles)", Useful Insects of the World series 3, I feel the lack of pictures of more species due to the difficulty of probe and collection of bugs. I will reflect it at the next book.

I would like to thank president and persons of Buan-gun, Dr. Seo Jieun and Dr. Oh Sunghwan (editorial supervisors), Professor Park Yunjeom and Professor Oh Sunhwan (advisors), Kim Taewan (the chief of Insect Research Institute), Park Cheolyoung (the representative of Communication Yeollim Co. that performed graphic design and publications of my books), staffs of Communication Yeollim Co. and especially Lee Sujin who performed hard composition and trimming of Insect pictures. And, I would like to thank all the other persons who helped to write and release the books. Finally, I would like to thank my mother who gave me constant energy, the late father, and my sons Changhyeon and Changgeun who did not complain of my late coming home with the excuse of writing and encouraged me.

April 2009    Author **Son Minwoo**

# 발간사

"학문하는 사람은 정신을 가다듬어 한 곳에 집중해야 한다. 만일 덕을 닦으면서도 마음을 공적(功績)과 명예에 둔다면 틀림없이 깊은 경지에까지 이르지는 못할 것이며, 책을 읽으면서도 읊조리는 맛이나 놀이에만 감흥을 느낀다면 결코 깊은 마음에 이르지 못할 것이다." – 채근담 –

위의 문구를 도감제작에 임하는 기본정신이라 생각하며 마음의 끈을 놓지 않으려고 제작기간 동안 항상 노심초사해 왔습니다. 후예들에게 소중한 유산으로 물려줄 수 있는 그런 책을 한번 만들어 보겠다고...

지금으로 부터 5년 전 걸려온 한통의 전화.
"지금 귀사에서 제작하신 명함을 보고 있는데 이정도의 퀄리티로 책도 제작이 가능합니까?"
"예, 당연히 제작이 가능 합니다."
이렇게 명함 한 장이 인연의 고리가 되어 손민우 선생과의 만남이 시작되었습니다. 처음 저자가 직접 촬영, 합성한 초정밀 고해상도 곤충사진(1편의 표지 모델인 루더킹멋쟁이사슴벌레)을 본 순간 "이정도의 멋진 원고라면 그동안 쌓아온 우리의 기술기반을 토대로 세계 최고의 도감을 만들 수 있겠다"고 확신하게 되었고, 그로인해 도감전문 출판사로의 첫발을 내딛게 되었습니다.

예상을 못했던 것은 아니지만 도감을 출판하는 일은 카탈로그 등의 단종 인쇄물을 만드는 것과는 차원이 다른 작업이었습니다. 작은 곤충을 크게 확대하여 표현하기 위해 원고의 미세한 편차도 용납할 수 없었으며, 곤충이 갖고 있는 아름다운 색상까지도 자연의 색상에 가깝게 구현해야 했기에 일반 인쇄기술 수준을 뛰어넘는 첨단의 편집과 인쇄기법을 동원하여야만 하였습니다. 진행되는 매순간마다 헤아릴 수 없이 많고 다양한 시행착오를 겪으며 지쳐 쓰러질 정도로 힘들었으나 저자의 식지 않는 열정과 출판사의 자존심을 건 실험정신이 토대가 되어, 2년여라는 결코 짧지 않은 세월을 거치면서 1편 "세계의 사슴벌레 대도감"이 세상에 나오게 되었던 것입니다. 1편 도감을 제작, 완성할 즈음에는 가히 "등골이 휠 것 같다"는 말이 떠오를 정도로 혼신의 힘을 다 소진한 느낌이었습니다.

도감을 제작해보니 도감은 좋은 표본만 많이 갖고 있다고 해서 만들 수 있는 것도 아니요, 학식이 높고 기술이 뛰어나다고 해서 만들어질 수 있는 것도 아니라는 것을 깨닫게 되었습니다. 집요한 열정과 의지 그리고 그 바탕에 진지하게 깔려 있는 바른 인생관이 초석이 되어야 한다고 생각됩니다.

이제 2편 "세계의 장수풍뎅이 대도감"과 3편 "세계의 꽃무지 대도감"을 연이어 출간하게 되니 도감시리즈의 모양새가 좀 더 갖추어져 기쁨과 안도의 마음이 앞서는 것이 사실입니다. 하지만, 앞으로 더욱 더 펼쳐질 초정밀 도감제작과 많은 응용 아이템을 앞에 두고 "시작이 반"이라는 말 대신 "반부터 새로운 시작"이라는 생각으로 마음의 잣대를 다시 세워봅니다.

마지막으로 열혈 학자이신 손민우 선생과 도감이 출간되기까지 도움을 주신 많은 분들께 깊은 감사를 드립니다. 이러한 열정과 도움들이 머지않은 미래에 큰 기쁨이 될 수 있도록 도감의 시리즈가 완성되는 그날 까지 최신 트렌드와 기술을 적극적으로 도입하여 세계시장에 한국출판의 위상을 높일 수 있는 명실공이 세계 최고 수준의 도감으로 빚어낼 것입니다.

2009. 4.   (주)커뮤니케이션열림 대표 **박철영**

# Address of Publication

"Those who pursue learning should pull themselves together and focus on one thing. If you cultivate virtue and have your heart set on credit and honor at the same time, You would never reach the ultimate stage. If you read a book but are inspired only by the chanting and playing, it will never reach your heart."   - Chaegeundam -

Thinking of the words above as the basic principle for publishing the pictorial books, I have constantly tried not to loosen my heart up. I wanted to make a book that I could bequeath to descendents as precious heritage.

I had a call five years ago.
"I am looking at the business card that you made. Could you make a book of this quality?"
"Yes, of course we could."
With one piece of business card as the medium of relationship, Son Minwoo and I met each other first. Upon first seeing the high precision and resolution insect photos(*Odontolabis ludekingi*, the cover page model of Volume 1) that the author took pictures of, I took the first step to work for the publishing company of the pictorial book, certain that with such a nice draft, we could make the best pictorial book in the world based on the technical skills we had accumulated.

Although it was not very different from what I expected, publishing pictorial books were totally different from making simple printed matters such as catalogue. Even subtle difference in the draft was not permitted in magnifying and expressing small insects, and the beautiful colors of insects should be represented just as the original colors. Thus, cutting-edge editing and printing techniques beyond the normal printing skills were required. Every moment that I went through numerous trials and errors, I was so exhausted that I almost knocked down, but the un-diminishing passion of the author and the spirit of challenge and pride have led to publishing the 1 volume of "The Lucanid Beetles of the World." more than 2 years after its beginning. As the first volume was almost completed, I felt like putting every ounce of my energy to the point that the saying, "the back is breaking" came up to my mind.

While making this pictorial book, I realized that pictorial books require more than mere many good samples, scholarly attainment, and excellent technology. I am certain that consistent passion, will, and upright view of life as the basis should be the footstone for it.

Releasing the 2nd volume, "The Dynastid Beetles of the World," and the 3rd volume, "The Cetoniid Beetles of the World," I truly felt joy and relief since the series started getting into shape. However, thinking of high precision pictorial books and various other items to be produced in the future, I once again recovered my composure, thinking that the second half is a new beginning, rather than well begun is half done.

Lastly, I deeply appreciate those who have helped me including Son Minwoo, a passionate scholar, to the point of publishing the pictorial book. In order for such passion and help to produce great joy in the near future, we will actively introduce up-to-date trends and technologies so that this pictorial book can be recognized as enhancing the nation's status regarding publication in world market in name and reality.

April 2009     Representative of Communication Yeollim   **Park Cheolyoung**

# 추천사

"세계의 사슴벌레 대도감"이 출간된 지 2년 여 만에 후속 시리즈인 "세계의 장수풍뎅이 대도감"과 "세계의 꽃무지 대도감"이 출간되었다는 소식은 본인 뿐만 아니라 곤충을 연구하는 많은 분들을 자랑스럽게 만들기에 충분할 뿐 아니라 단순의미 이상의 기쁨을 선사하게 될 것입니다.

손민우 선생을 알고 지내온 지난 수 년 여의 주마등 같은 시간, 5년 전 어느 날, 손민우 선생이 우리 연구소를 방문 했을 때 처음 풀어 놓은 거짓말 같은 얘기, 세계의 곤충 대도감 시리즈를 기획하고 있다는 소식을 접했을 때 저는 내심 반가웠음에도 걱정부터 앞섰던 것이 사실입니다.

그러나 지칠 줄 모르는 집중력으로 이루어 가는 그 분의 작업을 지켜보면서 제 염려는 무지에서 비롯된 기우였음을 깨닫게 되었습니다. 넘치는 열정뿐 아니라 진보된 표본촬영기술과 모든 첨단 인쇄기술이 동원된 고해상도의 곤충 대도감이 연작으로 출간되는 것을 결국 지켜보게 되었습니다. 이는 실로 한국 출판문화의 쾌거가 아닐 수 없습니다. 앞으로 세계 시장에 보급과 아울러 한국의 위상을 전할 수 있는 훌륭한 매체의 이정표가 되리라 의심치 않습니다. 한국과 세계 곤충 마니아들의 많은 관심과 성원을 부탁드리며, 앞으로 출간될 시리즈의 완간을 기대하면서 추천사를 갈음하고자 합니다.

2009. 4.   전 국제곤충연구소 회장 **김태완**

The news that "The Dynastid Beetles of the World" and "The Cetoniid Beetles of the World" were published two years after "The Stag Beetles of the World" was issued as the following series will make all insect researchers including me proud as well as present meaningful joy to them.

The memories for the last years since I first met Son Minwoo go through my mind like a flash. About five years ago when he visited our research center, I was glad and at the same time a bit worried when I heard the surprising news that he was planning to publish Insects book of the World series.

However, as I observed him achieving things with unflagging concentration, I realized that my worry was out of ignorance and groundless. Eventually, his burning passion, advanced sample filming technology, and cutting-edge printing technology made it possible to publish the high resolution Insects book of the World Series. This is indeed a brilliant achievement in the publication culture of Korea. Without a doubt, this will be the milestone of advancement into world markets and enhancing the position of Korea. Looking forward to completing the publication of the following series, I hope that enthusiasts around the world give continued interest and support for this project.

April 2009   Former president of International Insect Research Center **Kim Taewan**

우리나라 최초의 본격적인 세계 갑충도감인 "세계의 사슴벌레 대도감"에 이어 후속으로 "세계의 장수풍뎅이 대도감"과 "세계의 꽃무지 대도감"의 출간을 진심으로 축하드립니다.

앞서 많은 선진국에서 곤충도감이 출간되어 왔지만 본 곤충도감 시리즈는 도판의 세밀함과 정밀함이 다른 도감들과 차별하기에 충분해 보이고 또한 돋보기로 관찰하듯 곤충의 미세한 부분까지도 나안(裸眼)으로 볼 수 있으며 살아서 튀어나올 듯한 생동감은 본 도감이 지닌 독보적인 기술이라 칭송하고 싶습니다. 더불어 분류체계에 따라 정보를 잘 정리하여 독자들이 한 눈에 각각의 종들을 이해하고 분류동정하는데 손쉬울 수 있도록 설명함 역시, 이 도감의 가치를 빛나게 하는 힘이 되고 있습니다.

꿈을 위해 돌탑을 쌓듯, 한 편 한 편 도감을 완성해 가는 손민우 선생의 "세계유용 곤충대도감시리즈"가 곤충을 사랑하고 좋아하는 모든 동호인들에게 사랑을 받게 될 것이라 확신합니다. 나아가 그 열정과 독보적인 지식과 기술력을 바탕으로 전 세계에 한국의 자긍심을 높이는 주춧돌이 되어 주시길 기원해 봅니다.

2009. 4.   전 나비학회장 농학박사 **오성환**

I heartily congratulate the publication of "The Dynastid Beetles of the World" and "The Cetoniid Beetles of the World", the following series of "The Lucanid Beetles of the World" which was the first full-out pictorial book series of the beetle in Korea.

Although many advanced countries have published pictorial books of insects, but this pictorial book series are far superior from them with it's minuteness and high density in detail. Even minute areas of insects are observable with naked eyes as if you observe them with spectacles. Especially, I would like to commend the matchless technology that makes the pictures so vivid that they would come to life. In addition, the arrangement of information and classification to considered the reader comfort also enhances the values of this book.

I am sure that "Useful Insects of the World series" of Son Minwoo, who completes one book after another as if he builds up a stone tower for a dream, will be loved by all who like insects. Further, may this series be a milestone to enhance the pride of Korea over the world based on the passion and unique technology.

April 2009   Former president of Korea Butterfly Society **Oh Sunghwan**

# Recommendation

지금 박사학위 과정을 밟고 있는 손민우 선생을 만난 것은 그가 곤충대도감 시리즈를 기획 하면서 부터이다. "세계의 사슴벌레 대도감"에 이어 "세계의 장수풍뎅이 대도감"과 "세계의 꽃무지 대도감"이란 주제에서 보듯 손민우 선생은 언제나 "세계"라는 단어를 맨 앞에 붙이곤 하였다. 그 세계라는 의미가 많은 것을 내포 하고 있겠지만 디자인과 조형을 공부한 나로서는 "그렇다면 세계 속에 대한민국의 자존심을 걸고 한번 만들어 보자"는 주문을 걸고 시간 날 때마다 도감의 편집 자료를 가지고 허심탄회하게 이야기를 나누었던 기억이 새롭다.

세상의 만물이 그렇듯, 그 어떤 것이 완벽하게 꽃을 피울 수 있다는 것은 그 존재 자체의 끝없는 겸손과 낮춤, 시련을 이겨 낼 수 있는 인내와 용기, 그리고 내재된 열정과 집중력을 가지고 뜻 한바 목적을 향해 부단히 노력하지 않으면 이루어 질 수 없는 것이라 본다.

그 표본이 바로 손민우 선생이 40여 년을 잉태해 내 놓는 세계 장수풍뎅이와 꽃무지 도감이다. 이제 손민우 선생의 노고를 치하하기 보다는 본인의 뜨거운 탐구열과 하고자 하는 어떤 사명감에 오직 그것만을 위해 미쳐 살아가는 그 열정에 깊이 머리 숙인다

출판된 도감을 보면서 21세기 첨단 과학의 위력에 감사 한다. 뿐만 아니라 제작 관리에 아낌없는 혼열을 쏟아 부은 (주)커뮤니케이션열림의 박철영 사장님께 존경을 표한다. 왜냐하면 실물보다 더 아름답고 격조 높은 감동을 주는 그림 같은 도감을 만들어 주셨기 때문이다.

이제 방학이 되면 이 도감을 가지고 로마와 독일에 가서 옛 친구들에게 보여주고 한껏 자랑 하고 싶다. 코리아에도 손민우 선생 같은 분이 있고 이런 도감을 만들어 낼 수 있다고...

2009. 4.  전 한국조폐공사 디자인 실장 / 중부대학교 인쇄미디어학과 교수 **오순환**

I first met Son Minwoo, who is now working to get a doctorate, when he was planning the Insects in the World series. For all of his works - "The Lucanid Beetles of the World", "The Dynastid Beetles of the World" and "The Cetoniid Beetles of the World", he proudly put "the world" in the themes. The world may involve many meanings, but as for me, who studies design and modeling, come across the memories that we had frank conversation on the edition materials thinking, "let us make works with the pride of Korea in the world".

As everything does in the world, for something to be able to bloom completely, there should be consistent efforts to achieve goals as well as modesty and humility of being, patience and courage to face trials, and inherent passion and concentration.

The very representation of the principle is Son Minwoo, who issued "The Dynastid Beetles of the World" and "The Cetoniid Beetles of the World" after more than 40 year research. Now not merely commending the efforts that he has put forth, I would like to honor his burning passion and wholehearted attention to his commission.

I cannot but appreciate the power of the 21st century advanced science while looking through the published books. Besides, I extend my respect to Park Cheolyoung, the president of Communication Yeollim Co. who put every ounce of his energy for the production and management since he created the books delivering such a beautiful and lofty pictures and deep emotions.

I will go to Rome and Germany with this book during the vacation, and proudly show it to my old friends, saying that Korea has such prominent scholars as Son Minwoo who can publish this wonderful illustrated book.

April 2009

Former director of the Design Division, Korea Minting and Security Printing Corporation

Professor of the dept. of Printing Media, Joongbu University

**Oh Sunhwan**

# 차 례

| | |
|---|---|
| 4 | 머리말 |
| 6 | 발간사 |
| 8 | 추천사 |
| 10 | 차례 |
| 12 | 목차 |
| 14 | 명칭설명 |
| 16 | 본문페이지구성방식 |
| 18 | 곤충의지리적구분 |
| 19 | 곤충의주요채집국 |
| 20 | 서문 |
| 22 | 모양찾기 |
| 36 | 본문 |
| 164 | 도감의특징 |
| 165 | 사진제작과정 |
| 166 | 저자소개 |
| 168 | 에피소드 |
| 178 | 도움주신분들 |
| 179 | 참고문헌 |
| 180 | 부안누에타운 곤충과학관 |
| 181 | 곤충전시시설 |
| 182 | 찾아보기 |

# Contents

| | |
|---|---|
| 5 | Preface |
| 7 | Introduction |
| 8 | Recommendation |
| 11 | Contents |
| 12 | Table of Contents |
| 14 | Terms |
| 17 | How this Book Works |
| 18 | Geographical Classification of Insect |
| 19 | Major Countries where Insects are Collected |
| 21 | Foreword |
| 22 | Identification Key |
| 36 | Main Pages |
| 164 | Special Point of this Book |
| 165 | Photo Produce Process |
| 166 | About the Author |
| 168 | Episode |
| 178 | Thanks to |
| 179 | References |
| 180 | Buan Silkworm Town Insect Science Museum and Insect Ecology Exploration Science Park |
| 181 | Insect Exhibition Facility |
| 182 | Index |

# 목 차

## Section 1

| | |
|---|---|
| 개미집살이꽃무지족 Cremastocheilini | Tribe 1 |
| 거짓개미집살이꽃무지족 Xiphoscelidini | Tribe 2 |
| 마다가스카르꽃무지족 Stenotarsiini | Tribe 3 |
| 오스트레일리아꽃무지족 Schizorhinini | Tribe 4 |

| | | | | |
|---|---|---|---|---|
| 노랑마다가스카르꽃무지 *Euchroea auripimenta* | 38 | 마크래이꽃무지 *Trichaulax macleayi* | 39 |

## Section 2

### 골리앗대왕꽃무지족 Goliathini — Tribe 5

| | | | | |
|---|---|---|---|---|
| 골리앗레기우스대왕꽃무지 *Goliathus regius* | 42 | 크라치지줄무늬귀신꽃무지 *Mecynorhina kraatzi* | 82 |
| 골리앗대왕꽃무지 *Goliathus Goliathus* | 44 | 사바게이점박이귀신꽃무지 *Mecynorhina savagei* | 83 |
| 골리앗대왕꽃무지 (아피카리스 형) *Goliathus goliathus*-Form *apicalis* | 46 | 주홍대왕귀신꽃무지 (유니칼라 형) *Mecynorhina oberthueri*-Form *unicolor* | 84 |
| 골리앗대왕꽃무지 (콘스페르서스 형) *Goliathus goliathus*-Form *conspersus* | 48 | 주홍점박이대왕귀신꽃무지 (데코라타 형) *Mecynorhina oberthueri*-Form *decorata* | 85 |
| 골리앗대왕꽃무지 (비타투스 형) *Goliathus goliathus*-Form *vittatus* | 50 | 토르콰타-이마쿠리콜리스대왕귀신꽃무지 *Mecynorhina torquata immaculicollis* | 86 |
| 골리앗대왕흰꽃무지 (푸스투라투스 형) *Goliathus orientalis*-Form *pustulatus* | 51 | 토르콰타-포게이대왕귀신꽃무지 *Mecynorhina torquata poggei* | 87 |
| 골리앗오리엔탈흰대왕꽃무지 *Goliathus orientalis* | 52 | 토르콰타-우간덴시쓰대왕귀신꽃무지 *Mecynorhina torquata ugandensis* | 88 |
| 골리앗흰대왕꽃무지 (프레이시 형) *Goliathus orientalis*-Form *preissi* | 54 | 하리시큰뿔꽃무지 *Megalorhina harrisi harrisihi* | 92 |
| 골리앗흰대왕꽃무지 (운두라투스 형) *Goliathus orientalis*-Form *undulatus* | 56 | 하리사-엑스미아큰뿔꽃무지 *Megalorhina harrisi eximia* | 93 |
| 골리앗알보씨그나투스-키르키아누스대왕꽃무지 *Goliathus albosignatus kirkianus* | 58 | 로우예리쌍뿔꽃무지 *Mystroceros rouyeri* | 94 |
| 골리앗카시쿠스대왕꽃무지 *Goliathus cacicus* | 60 | 아우조욱시뿔꽃무지 *Neophaedimus auzouxi* | 95 |
| 포르나시니왕꽃무지 *Fornasinius fornasinii* | 62 | 스탄레이넵튠꽃무지 *Neptunides stanleyi* | 96 |
| 루쑤스왕꽃무지 *Fornasinius russus* | 63 | 폴리크로우스넵튠꽃무지 *Neptunides polychrous polychrous* | 97 |
| 코로사얼룩뿔꽃무지 *Hypselogenia corrosa* | 64 | 폴리크로우스넵튠꽃무지 *Neptunides polychrous* | 98 |
| 왈라치사슴풍뎅이 *Dicranocephalus wallichii* | 65 | 베르토니흰큰머리꽃무지 *Rhamphorrhina bertolonii* | 100 |
| 사슴풍뎅이 *Dicarnocephalus adamsi* | 66 | 스프레덴스흰큰머리꽃무지 *Rhamphorrhina splendens* | 101 |
| 부르케이앞장다리꽃무지 *Cheirolasia burkei burkei* | 68 | 풍이 *Pseudotorynorrhina japonica* | 102 |
| 부르케이셉텐트리오니스앞장다리꽃무지 *Cheirolasia burkei septentrionis* | 69 | 레스프렌덴스금광풍뎅이 *Rhomborhina resplendens* | 103 |
| 데르비아나왕꽃무지 *Dicronorrhina derbyana derbyana* | 70 | 구타타기린뿔꽃무지 *Stephanorrhina guttata* | 104 |
| 데르비아나콘라드씨왕꽃무지 *Dicronorrhina derbyana conradsi* | 71 | 줄리아기린뿔꽃무지 *Stephanorrhina julia* | 105 |
| 데르비아나오벨투에리왕꽃무지 *Dicronorrhina derbyana oberthueri* | 72 | 롱기셉스앞뿔꽃무지 *Taurrhina longiceps* | 106 |
| 큐프레오수투랄리스뿔꽃무지 *Eudicella cupreosuturalis* | 73 | 셉타노랑띠꽃무지 *Pedinorrhina septa* | 107 |
| 그랄리뿔꽃무지 *Eudicella gralli gralli* | 74 | 알페스트리스꽃무지 *Tmesorrhina alpestris* | 108 |
| 그랄리움부로비타타뿔꽃무지 *Eudicella gralli umbrovittata* | 75 | 알페스트리-바후텐시스꽃무지 *Tmesorrhina alpestris bafutensis* | 109 |
| 스미씨쉬라티카뿔꽃무지 *Eudicella smithi shiratica* | 76 | 이리스긴몸광꽃무지 *Tmesorrhina iris* | 110 |
| 트릴리네아뿔꽃무지 *Eudicella trilineata* | 77 | 라에타연보석꽃무지 *Tmesorrhina laeta* | 111 |
| 유리히나붉은다리주걱턱꽃무지 *Ingrisma euryrrhina* | 78 | 로칠디주걱턱꽃무지 *Trigonophorus rothschildi* | 112 |
| 오리바세우스사슴뿔꽃무지 *Cyphonocephalus olivaceus* | 79 | 후라메아풍이 *Torynorrhina flammea flammea* | 113 |
| 훼레로이미네티큐앞장다리꽃무지 *Jumnos ferreroiminettiique* | 80 | 후라메아-키케리풍이 *Torynorrhina flammea chicheryi* | 116 |
| 룩케리노랑네점박이앞장다리꽃무지 *Jumnos ruckeri* | 81 | | |

# Table of Contents

## Section 3
### 꽃무지족 Cetoniini — Tribe 6

| | |
|---|---|
| 네점박이붉은띠꽃무지 *Glycyphana sp.* … 120 | 괌꽃무지 *Protaetia guam* … 130 |
| 에피피아타-활케이주홍테꽃무지 *Pachnoda ephippiata falkei* … 121 | 오리엔탈리스점박이꽃무지 *Protaetia orientalis* … 131 |
| 에피피아타-환코이시주홍테꽃무지 *Pachnoda ephippiata francoisi* … 122 | 흰점박이꽃무지 *Protaetia brevitarsis seulensis* … 132 |
| 마르기나타주홍테꽃무지 *Pachnoda marginata* … 123 | 점박이꽃무지 *Protaetia orientalis submarmorea* … 133 |
| 쿠프레아점박이꽃무지 *Protaetia cuprea* … 124 | 필리핀점박이꽃무지 *Protaetia phillppensis* … 134 |
| 쿠프레아-올리바세아꽃무지 *Proteatia cuprea olivacea* … 125 | 스켑시아점박이꽃무지 *Protaetia scepsia* … 135 |
| 쎄레비카점박이꽃무지 *Protaetia celebica* … 126 | 우후리큰점박이꽃무지 *Protaetia uhligi* … 136 |
| 녹스점박이큰꽃무지 *Protaetia nox* … 127 | 베네라빌리스큰풀색꽃무지 *Protaetia venerabilis* … 137 |
| 루마위기점박이꽃무지 *Protaetia lumawigi* … 128 | 샤우미꽃무지 *Sternoplus schaumi* … 138 |
| 헝가리카꽃무지 *Protaetia hungarica* … 129 | |

## Section 4
### 모가슴꽃무지족 Gymnetini — Tribe 7

| | |
|---|---|
| 남미점박이모가슴꽃무지 *Gymnetis pantherina* … 142 | 지카니남미점박이모가슴꽃무지 *Gymnetis pantherina zikani* … 143 |

## Section 5
### 얼룩꽃무지족 Diplognathini — Tribe 8
### 투구꽃무지족 Phaedimini — Tribe 9

| | |
|---|---|
| 르히노필루스투구꽃무지 *Mycteristes rhinophyllus* … 146 | 볼렌호베니뾰족투구꽃무지 *Mycteristes vollenhoveni* … 149 |
| 티베타나두뿔투구꽃무지 *Mycteristes tibetana* … 147 | 하우드니투구꽃무지 *Phaedimus howdeni* … 150 |
| 스쿠아모수스뾰족투구꽃무지 *Mycteristes squamosus* … 148 | |

## Section 6
### 홀쭉꽃무지족 Taenioderini — Tribe 10

| | |
|---|---|
| 얼룩홀쭉꽃무지 *Euselates sp.* … 154 | 솔로모니카주홍줄홀쭉꽃무지 *Ixorida solomonica* … 159 |
| 스틱티카홀쭉꽃무지 *Euselates stictica* … 155 | 프로핀쿠아흰점홀쭉꽃무지 *Ixorida propinqua* … 160 |
| 엘레강스점박이홀쭉꽃무지 *Ixorida elegans* … 156 | 앤드로얘디홀쭉꽃무지 *Plectrone endroedii* … 161 |
| 후리데리씨금박무늬홀쭉꽃무지 *Ixorida friderici* … 157 | 루마위기검은박쥐꽃무지 *Plectrone lumawigi* … 162 |
| 아펠레스주홍줄홀쭉꽃무지 *Ixorida venerea apelles* … 158 | 트리칼라홀쭉꽃무지 *Taeniodera tricolor tricolor* … 163 |

## 명칭설명

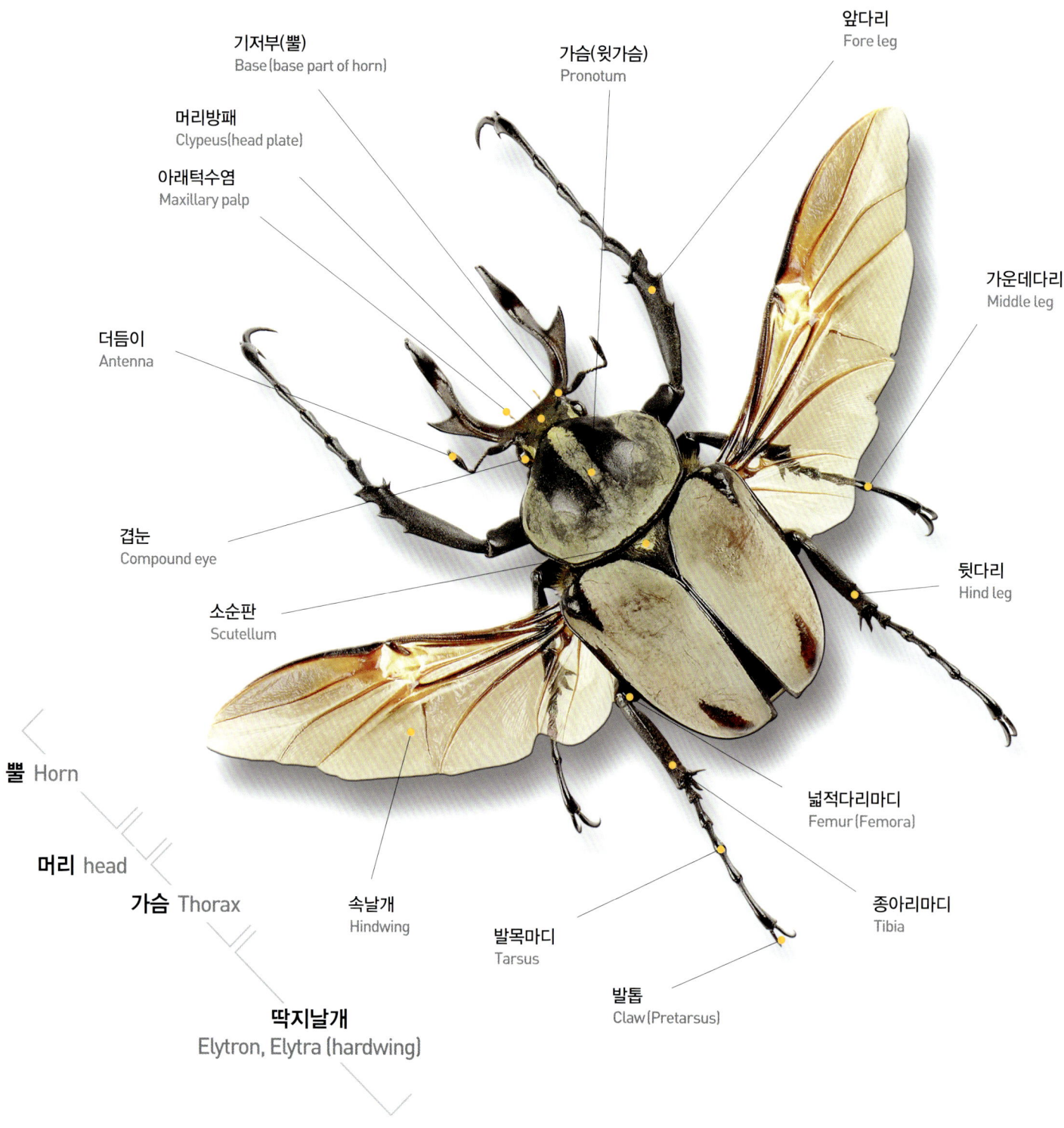

# Terms

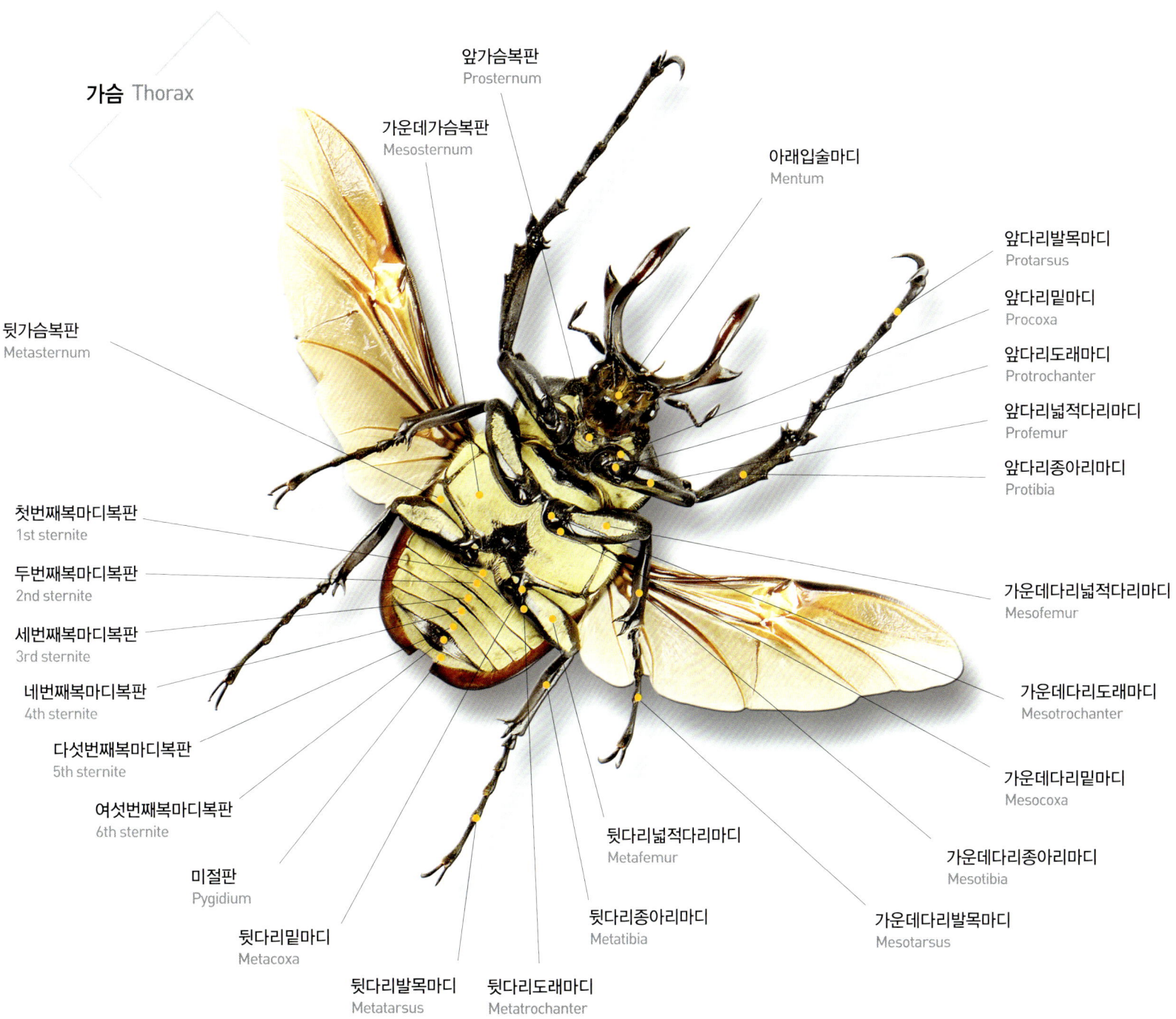

# 본문페이지구성방식

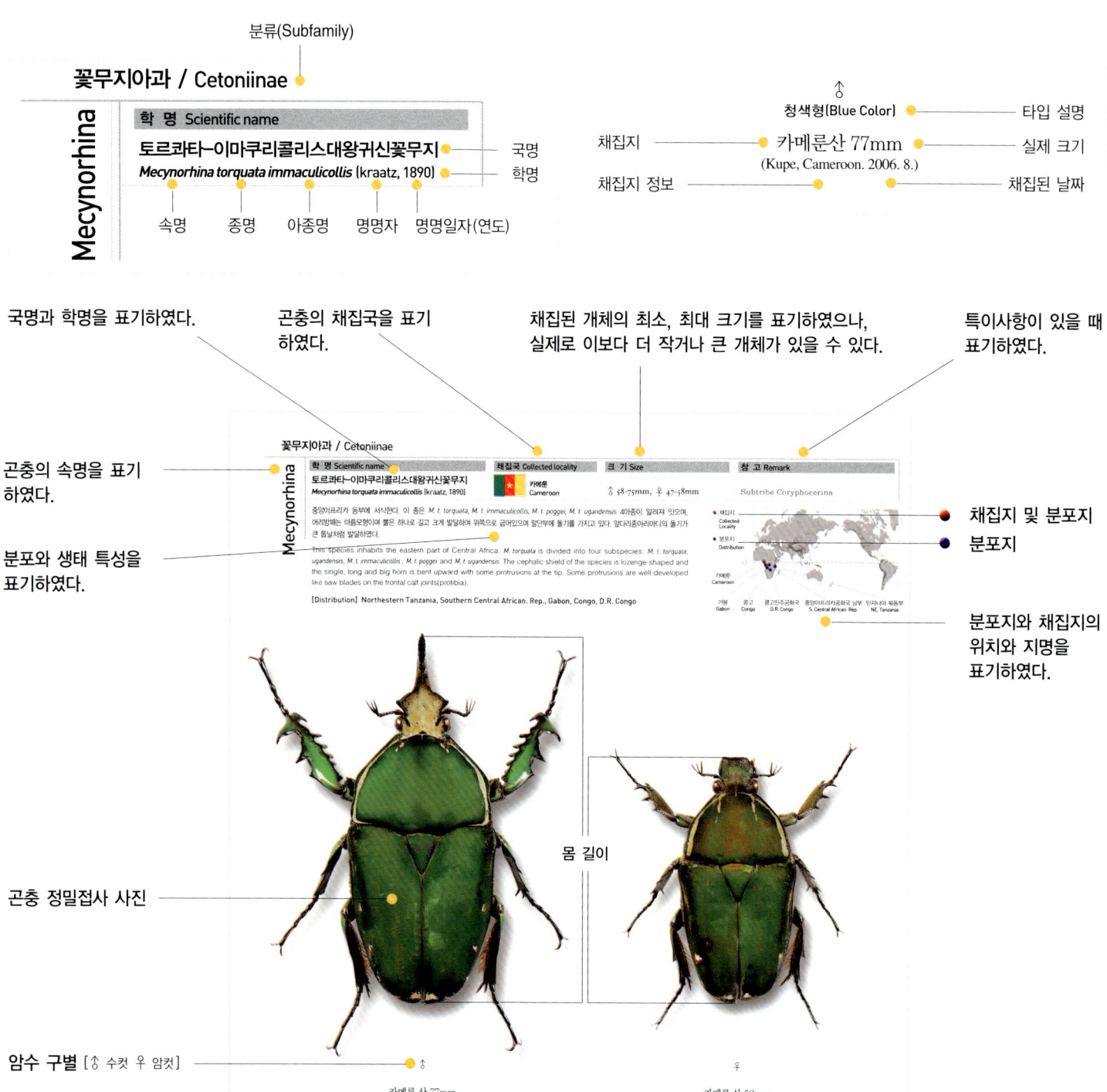

# How this Book Works

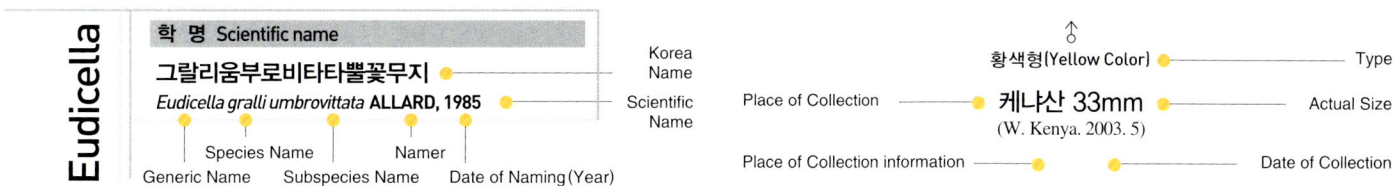

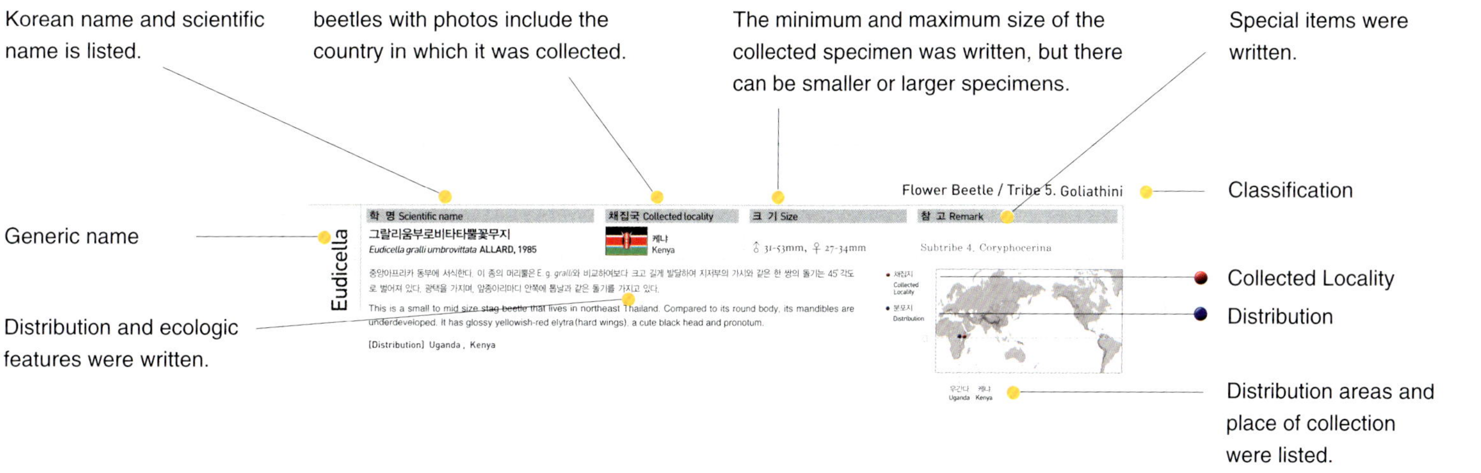

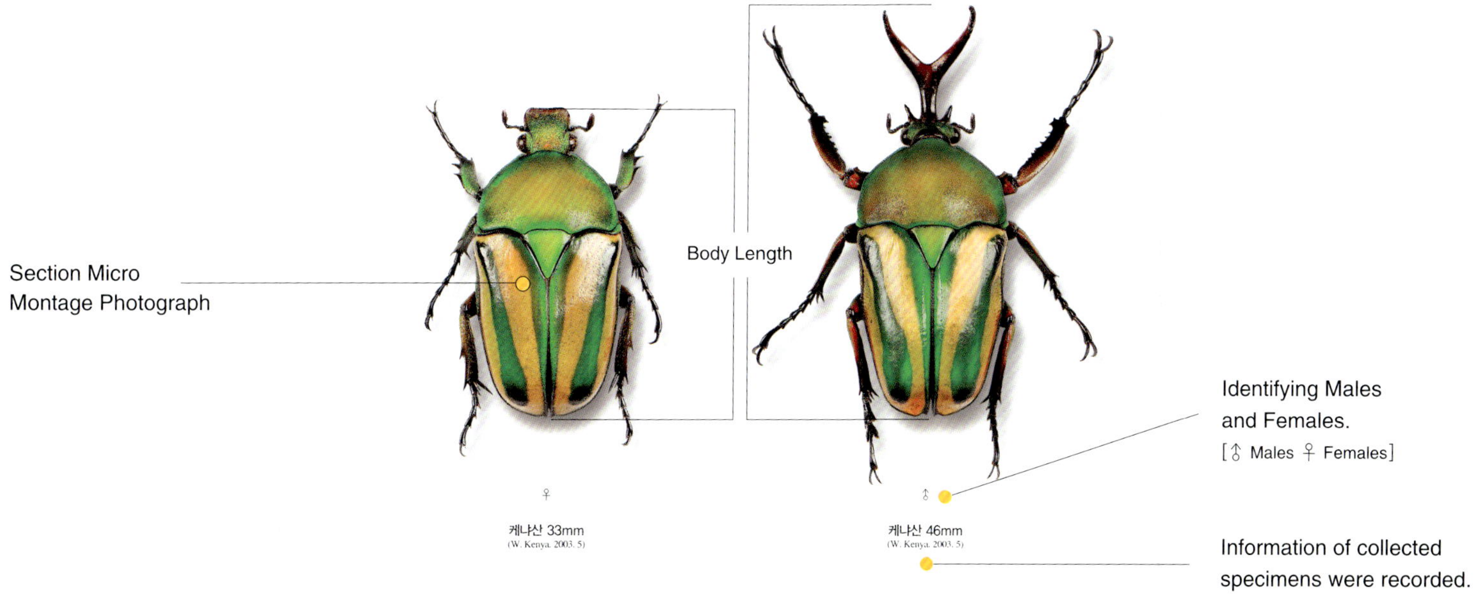

- Korean name and scientific name is listed.
- Generic name
- Distribution and ecologic features were written.
- beetles with photos include the country in which it was collected.
- The minimum and maximum size of the collected specimen was written, but there can be smaller or larger specimens.
- Special items were written.
- Classification
- Collected Locality
- Distribution
- Distribution areas and place of collection were listed.
- Section Micro Montage Photograph
- Identifying Males and Females. [♂ Males ♀ Females]
- Information of collected specimens were recorded.

Pictures in this illustrated book focus on showing each insect specimen in detail. By adjusting the size of an inner specimen by taking into consideration the size of its outer counterpart, readers can guess how large or small an insect actually is. Also, in order to make the most of the high-resolution pictures, insect-morphological explanation is reduced to the minimum for the pictures to enable an easy morphological comparison. The other elements of the book, well arranged not to interfere with pictorial communication, are edited in a way that delivers all information about each specimen totally and efficiently.

# 곤충의 지리적 구분 Geographical Classification of Insect

우리나라는 중국 대륙과 시베리아와는 육지로 연결되어 있고, 일본과는 바다를 사이에 두고 떨어져 있다. 그렇지만, 이들 지역에 사는 곤충을 비교해 보면 비슷한 종류가 많다. 곤충을 조사하여 보면 그 지역에 공통된 경향을 볼 수 있으며, 이 경향을 알기 위하여 지리적 구분을 하고 있다. 그러나 이 경계는 확실한 것은 아니며, 어디까지나 경향이라는 점을 알아두어야 한다.

The Korean Peninsula is connected to the mainland of China and Siberia while looking toward Japan across the sea. A lot of insect groups, however, share each of these areas as their place of cohabitation. By examining insect groups, some biological or geographical commonness can be found and such commonness is the reason for geographical classification. It should be noted, however, that this classification is not definitive.

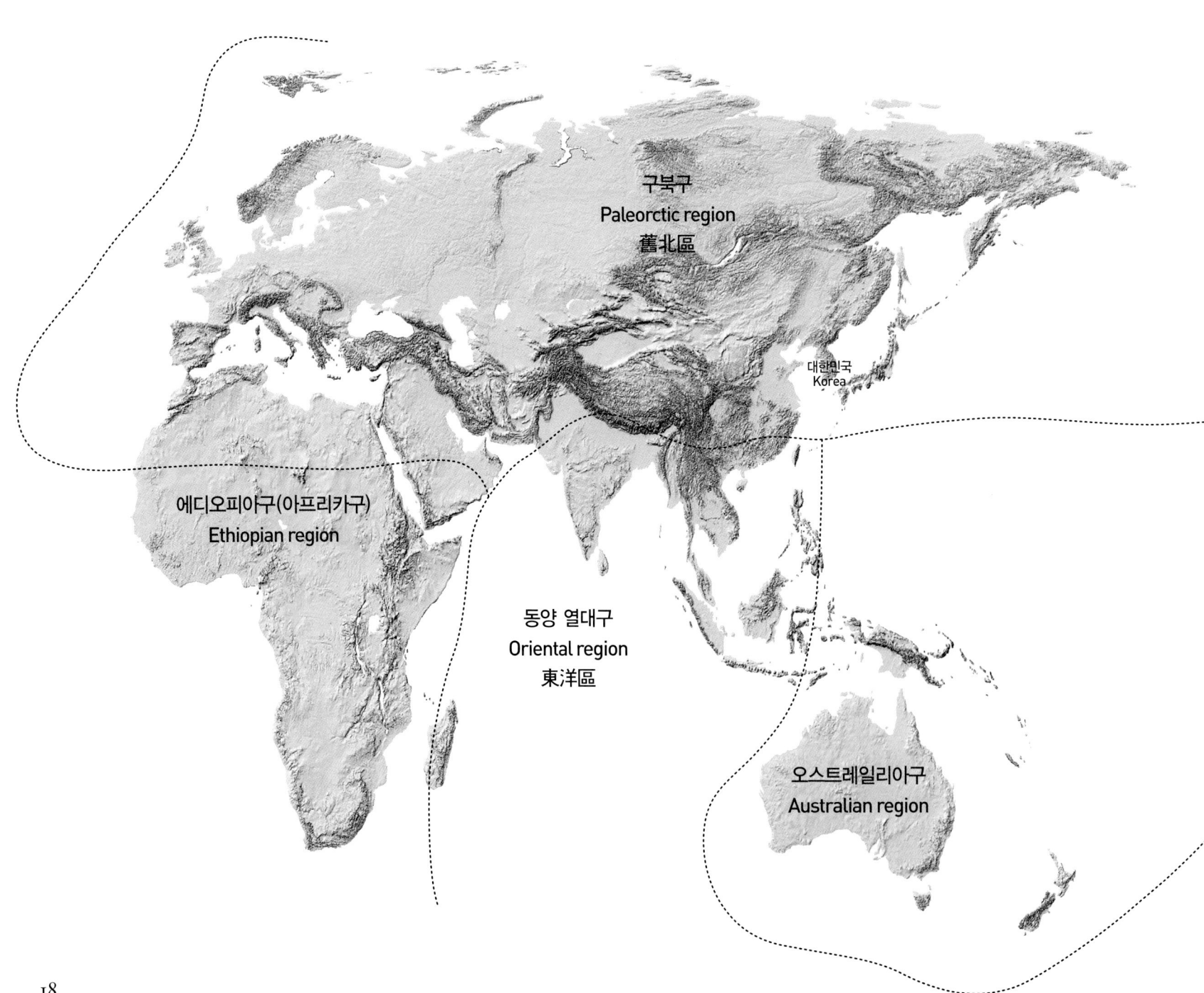

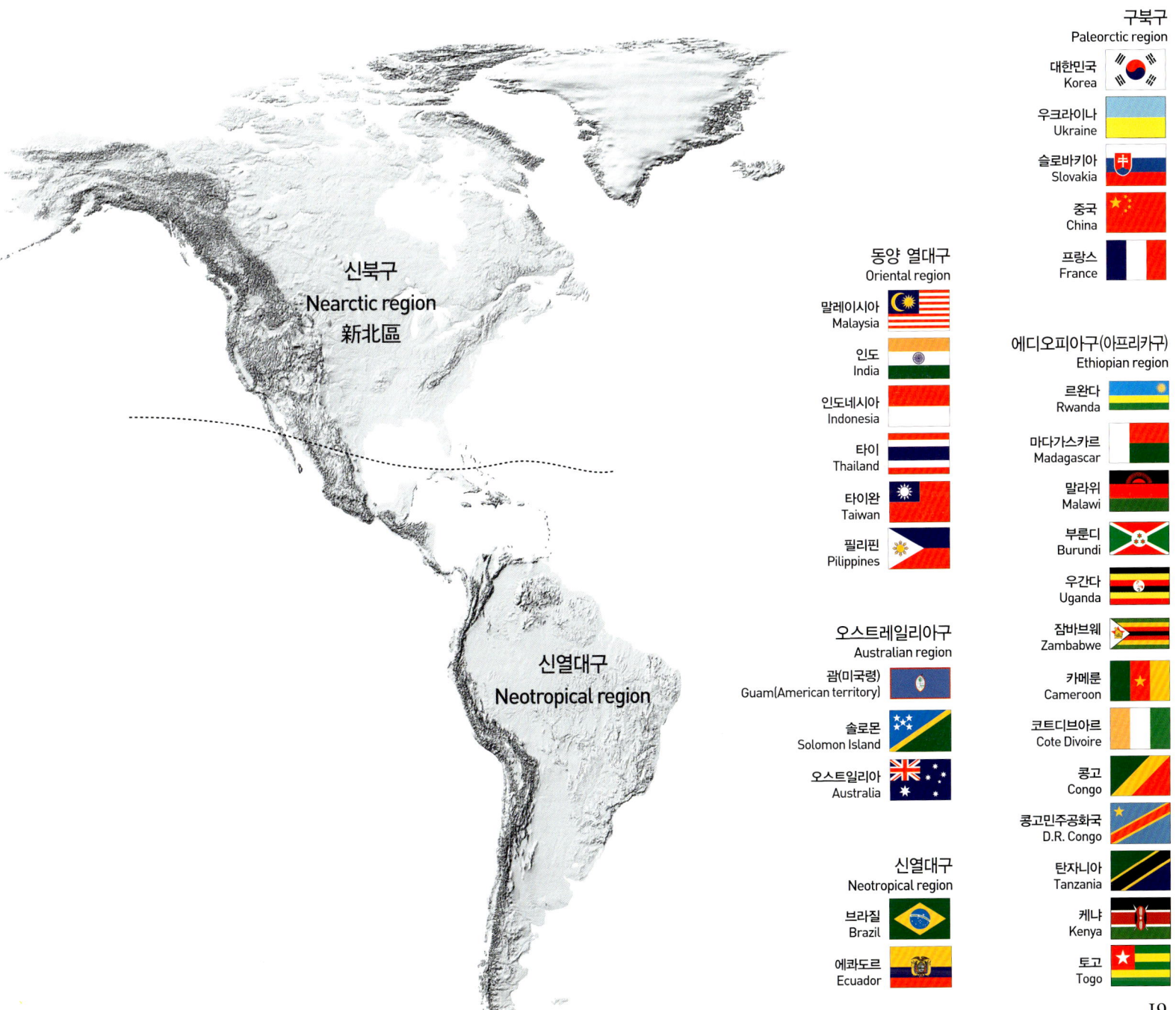

# 서문

꽃무지는(Cetoniid beetle)는 꽃에서 발견되기 때문에 영어로도 플라워 비틀(Flower beetle)이라고 부른다. 꽃무지는 꽃과 같이 색깔과 모양이 매우 다채로운 곤충이다. 현재까지 전 세계에는 약 3,200종의 꽃무지(Cetoniid beetles)가 알려져 있으며, 극지를 제외한 전 세계에 골고루 분포되어 있지만, 특히 아프리카와 동남아시아의 열대지역과 아열대지역에 약 70%가 서식하고 있다.

전 세계에 알려진 곤충은 약 100만~110만 종이 알려져 있으며, 해마다 새로운 종이 발견되고 있다. 그 중 꽃무지(Flower beetle)가 속해 있는 딱정벌레목 곤충은 가장 큰 분류군을 가지고 있으며, 약 40만종을 차지할 정도로 그 종류가 매우 많고 다양하다.

딱정벌레목의 특징은 두 쌍의 날개 중 초시(初翅, elytron)라 부르는 한쌍의 앞날개는 딱딱하게 굳어 있어 갑옷을 입은 것과 같이 몸을 보호할 수 있으며, 다른 한쌍의 뒷날개는 얇은 막 상으로 앞날개보다 크며, 비행시 사용하였다가 정지 시에는 앞날개 밑으로 접어 넣어 비행날개를 보호할 수 있는 보다 진보된 기능을 갖고 있다. 바로 이러한 점이 세계각지의 고산이나 평야, 하천과 늪, 지상이나 동굴, 식물체의 내부와 외부, 흙 속 등, 거의 모든 지역에서 적응하고 다양한 종으로 번성할 수 있게 되었다고 본다.

그 중 꽃무지는 딱정벌레목(Coleoptera) 풍뎅이과(Scarabaeidae) 꽃무지아과(Cetoniinae)에 속하며, 알-애벌레-번데기-어른벌레로 4단계의 성장과정을 거치는 완전변태 곤충으로 수명은 보통 1년이다.

우리나라의 꽃무지(Flower beetle)는 1994년도 한국곤충학회와 한국응용곤충학회에서 공동집필한 "한국곤충명집"에 넓적꽃무지아과(Valginae), 호랑꽃무지아과(Trichiinae), 꽃무지아과(Cetoniinae)로 3아과(Subfamily) 5족(Tribe) 16속(genus) 30종(species)으로 명기되어있었으나, 2002년도 농업과학기술원에서 발간한 "한국경제곤충" 풍뎅이상과 딱정벌레목편에 3아과(Subfamily) 1족(Tribe) 14속(genus) 19종(species)으로 동물이명(同物異名), 오동정, 기타의 오류 등으로 인하여 삭제되거나 개칭되었다. 마찬가지로 세계에 분포하는 꽃무지의 이름 또한 학자마다 분류의 기준이 다르기 때문에 많은 변동이 이루어지고 있으며, 본 도감에는 가장 최근의 종명을 따르고자 했다.

이 도감에는 우리나라의 꽃무지아과(Subfamily) 4속(Genus) 4종(Species) 7개체(Individuals)를 수록하였으며, 세계의 꽃무지는 꽃무지아과(Cetoniinae)에 해당하는 꽃무지만 나뉘어 39속(Genus), 97종(Species) 197마리의 초정밀 사진을 수록하였다.

# Foreword

Cetoniid beetle is called the Flower beetle in English as they are found on flowers. Like a variety of flowers, the Cetoniid beetle comes in various shapes and body colors. Approximately 3,200 species of Cetoniinae have been discovered all around the world except the north and south poles. Though their distribution is rather even worldwide, about 70% of them inhabit the African and Southeast Asian tropical and subtropical regions.

Currently, approximately 1 million to 1.1 million insect species are known worldwide with some new species discovered year by year. Coleoptera order, which includes Cetoniinae (also called flower beetle), is one of the biggest insect groups of them all with some 0.4 million species falling under it.

The order of Coleoptera is characterized by the front pair of wings, called elytron, being hard and functioning as a protective mechanism like a suit of armor while the rear pair, thin like films, is bigger than the other pair. The latter is used only when an insect in this order flies and kept under the front pair of wings for protection when it is not flying, which shows an advanced evolutionary function. This is seen as a reason this order can adapt to and diversify itself successfully in almost every ecosystem worldwide alpine area, plain, stream, wetland, cavern, inside and outside of plants, earth, etc.

Of the insect systematic categories, the Flower beetle is included in the order of Coleoptera, Scarabaeoidae family of subfamily of Cetoniinae. They go through a 4-step complete metamorphic process (egg - larva - pupa - adult) and their lifespan is about 1 year on average.

Previously, South Korean entomologists classified the Cetoniinae inhabiting the Korean peninsula into 30 species, 16 genera and 5 tribes under 3 subfamilies Valginae, Trichiinae and Cetoniinae according to the Check List of Insects from Korea, which was published in 1994 jointly by the Entomological Society of Korea and the Korean Society of Applied Entomology. The National Institute of Agricultural Science and Technology of South Korea, however, pointed out some errors in the previous classification, e.g. one species having two or more different names, and re-classified it into 19 Species, 14 genera and 1 Tribe under 3 subfamilies in its Korea Economy Insects published in 2002. Likewise, the names of the Flower beetles inhabiting other continents and regions are not established yet as there is still no consensus between entomologists regarding the matter. This book follows the most recent way of naming Flower beetles.

Introduced in this book are 7 individuals of the 4 genera and 4 Species of Cetoniinae living in South Korea. Also, the book carries some fine and detailed pictures of Cetoniid beetles, all of them in the subfamily of Cetoniid beetle and distributed worldwide. They are classified into 197 individuals of 39 genera and 97 Species in this book.

# 모양찾기

♂ 29mm
노랑마다가스카르꽃무지
*Euchroea auripimenta*
p. 38

♀ 32mm
노랑마다가스카르꽃무지
*Euchroea auripimenta*
p. 38

♂ 29mm
마크래이꽃무지
*Trichaulax macleayi*
p. 39

♀ 27mm
마크래이꽃무지
*Trichaulax macleayi*
p. 39

♂ 91mm
골리앗레기우스대왕꽃무지
*Goliathus regius*
p. 42

♂ 91mm
골리앗레기우스대왕꽃무지
*Goliathus regius*
p. 42

♂ 63mm
골리앗레기우스대왕꽃무지
*Goliathus regius*
p. 43

♂ 65mm
골리앗대왕꽃무지
*Goliathus Goliathus*
p. 45

♀ 76mm
골리앗대왕꽃무지
*Goliathus Goliathus*
p. 45

♂ 97mm
골리앗대왕꽃무지 (아피카리스 형)
*Goliathus goliathus* - Form *apicalis*
p. 46

# Identification Key

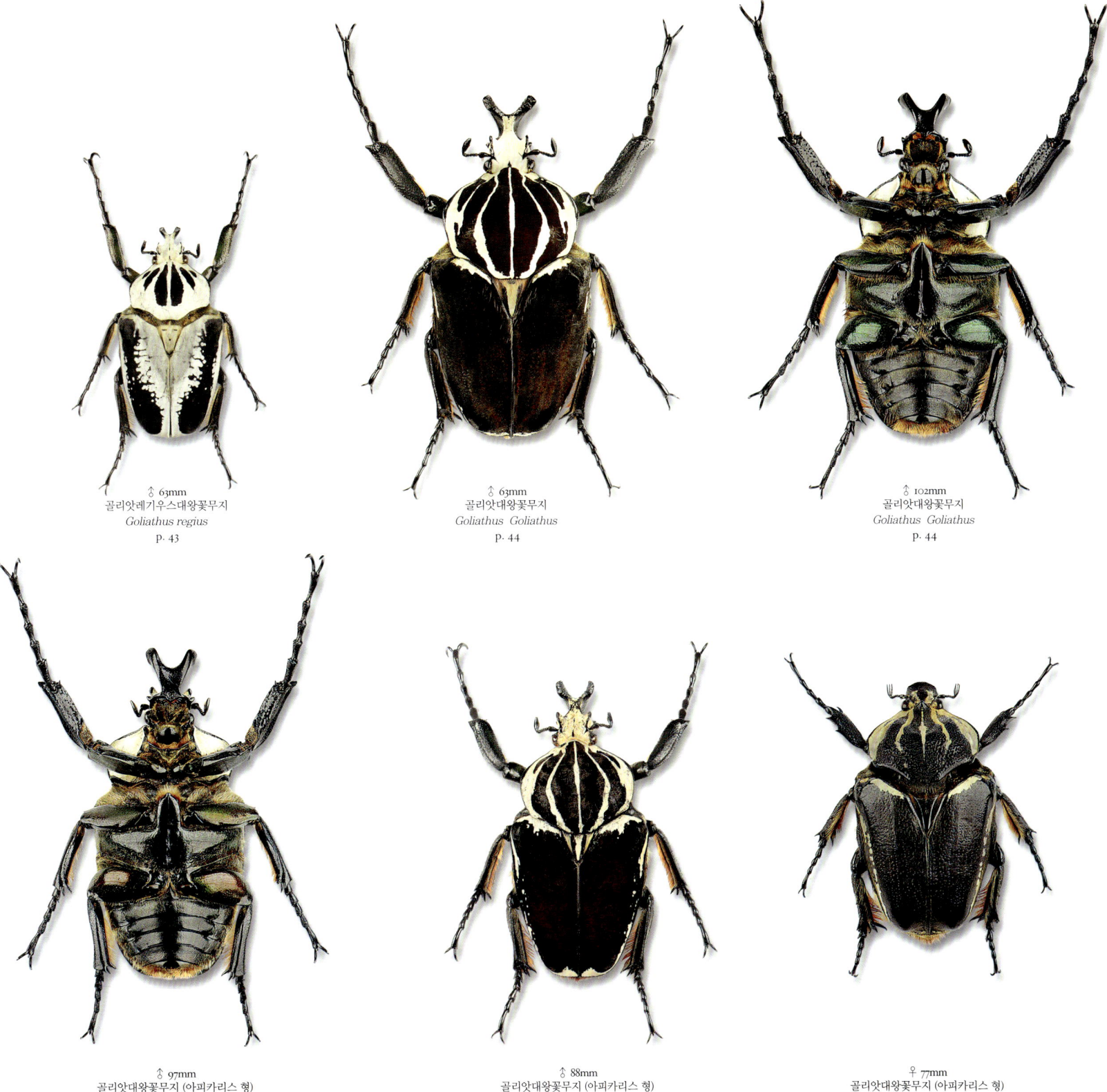

♂ 63mm
골리앗레기우스대왕꽃무지
*Goliathus regius*
p. 43

♂ 63mm
골리앗대왕꽃무지
*Goliathus Goliathus*
p. 44

♂ 102mm
골리앗대왕꽃무지
*Goliathus Goliathus*
p. 44

♂ 97mm
골리앗대왕꽃무지 (아피카리스 형)
*Goliathus goliathus* – Form *apicalis*
p. 46

♂ 88mm
골리앗대왕꽃무지 (아피카리스 형)
*Goliathus goliathus* – Form *apicalis*
p. 47

♀ 77mm
골리앗대왕꽃무지 (아피카리스 형)
*Goliathus goliathus* – Form *apicalis*
p. 47

# 모양찾기

0 10 20 30 40 50 mm

♂ 94mm
골리앗대왕꽃무지 (콘스페르서스 형)
*Goliathus goliathus* – Form *conspersus*
p. 48

♂ 94mm
골리앗대왕꽃무지 (콘스페르서스 형)
*Goliathus goliathus* – Form *conspersus*
p. 48

♀ 73mm
골리앗대왕꽃무지 (콘스페르서스 형)
*Goliathus goliathus* – Form *conspersus*
p. 49

♂ 90mm
골리앗오리엔탈흰대왕꽃무지
*Goliathus orientalis*
p. 52

♂ 90mm
골리앗오리엔탈흰대왕꽃무지
*Goliathus orientalis*
p. 52

♂ 57mm
골리앗오리엔탈흰대왕꽃무지
*Goliathus orientalis*
p. 53

# Identification Key

♀ 75mm
골리앗대왕꽃무지 (콘스페르서스 형)
*Goliathus goliathus* – Form *conspersus*
p. 49

♀ 45mm
골리앗대왕꽃무지 (비타투스 형)
*Goliathus goliathus* – Form *vittatus*
p. 50

♂ 82mm
골리앗대왕흰꽃무지 (푸스투라투스 형)
*Goliathus orientalis* – Form *pustulatus*
p. 51

♂ 78mm
골리앗대왕흰꽃무지 (푸스투라투스 형)
*Goliathus orientalis* – Form *pustulatus*
p. 51

♀ 70mm
골리앗오리엔탈흰대왕꽃무지
*Goliathus orientalis*
p. 53

♂ 75mm
골리앗흰대왕꽃무지 (프레이시 형)
*Goliathus orientalis* – Form *preissi*
p. 54

♂ 75mm
골리앗흰대왕꽃무지 (프레이시 형)
*Goliathus orientalis* – Form *preissi*
p. 54

# 모양찾기

♂ 59mm
골리앗흰대왕꽃무지 (프레이시 형)
*Goliathus orientalis* – Form *preissi*
p. 55

♀ 74mm
골리앗흰대왕꽃무지 (프레이시 형)
*Goliathus orientalis* – Form *preissi*
p. 55

♂ 87mm
골리앗흰대왕꽃무지 (운두라투스 형)
*Goliathus orientalis* – Form *undulatus*
p. 56

♂ 65mm
골리앗알보씨그나투스-키르키아누스대왕꽃무지
*Goliathus albosignatus kirkianus*
p. 58

♂ 58mm
골리앗알보씨그나투스-키르키아누스대왕꽃무지
*Goliathus albosignatus kirkianus*
p. 59

♀ 42mm
골리앗알보씨그나투스-키르키아누스대왕꽃무지
*Goliathus albosignatus kirkianus*
p. 59

♂ 79mm
골리앗카시쿠스대왕꽃무지
*Goliathus cacicus*
p. 60

# Identification Key

♂ 87mm
골리앗흰대왕꽃무지 (운두라투스 형)
*Goliathus orientalis* – Form *undulatus*
p. 56

♀ 62mm
골리앗흰대왕꽃무지 (운두라투스 형)
*Goliathus orientalis* – Form *undulatus*
p. 57

♂ 68mm
골리앗흰대왕꽃무지 (운두라투스 형)
*Goliathus orientalis* – Form *undulatus*
p. 57

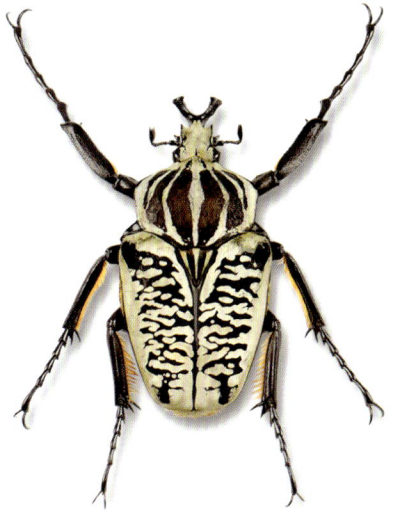

♂ 66mm
골리앗알보씨그나투스-키르키아누스대왕꽃무지
*Goliathus albosignatus kirkianus*
p. 58

♂ 79mm
골리앗카시쿠스대왕꽃무지
*Goliathus cacicus*
p. 60

♂ 67mm
골리앗카시쿠스대왕꽃무지
*Goliathus cacicus*
p. 61

♀ 70mm
골리앗카시쿠스대왕꽃무지
*Goliathus cacicus*
p. 61

♂ 48mm
포르나시니왕꽃무지
*Fornasinius fornasinii*
p. 62

♀ 47mm
포르나시니왕꽃무지
*Fornasinius fornasinii*
p. 62

# 모양찾기

♂ 59mm
루쓰스왕꽃무지
*Fornasinius russus*
p. 63

♀ 59mm
루쓰스왕꽃무지
*Fornasinius russus*
p. 63

♂ 25mm
코로사얼룩뿔꽃무지
*Hypselogenia corrosa*
p. 64

♀ 23mm
코로사얼룩뿔꽃무지
*Hypselogenia corrosa*
p. 64

♂ 36mm
왈라치사슴풍뎅이
*Dicranocephalus wallichii*
p. 65

♂ 27mm
사슴풍뎅이
*Dicarnocephalus adamsi*
p. 66

♀ 21mm
사슴풍뎅이
*Dicarnocephalus adamsi*
p. 66

♂ 38mm
데르비아나콘라드씨왕꽃무지
*Dicronorrhina derbyana conradsi*
p. 71

♀ 33mm
데르비아나콘라드씨왕꽃무지
*Dicronorrhina derbyana conradsi*
p. 71

♂ 44mm
데르비아나오벨투에리왕꽃무지
*Dicronorrhina derbyana oberthueri*
p. 72

♀ 39mm
데르비아나오벨투에리왕꽃무지
*Dicronorrhina derbyana oberthueri*
p. 72

♂ 33mm
큐프레오수투랄리스뿔꽃무지
*Eudicella cupreosuturalis*
p. 73

♂ 35mm
그랄리뿔꽃무지
*Eudicella gralli gralli*
p. 74

♀ 27mm
유리르히나붉은다리
주걱턱꽃무지
*Ingrisma euryrrhina*
p. 78

♀ 26mm
유리르히나붉은다리
주걱턱꽃무지
*Ingrisma euryrrhina*
p. 78

♂ 30mm
오리바세우스사슴뿔꽃무지
*Cyphonocephalus olivaceus*
p. 79

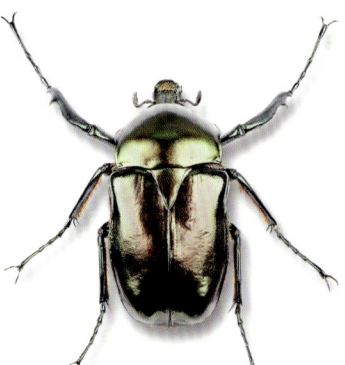

♂ 51mm
훼레로이미네티큐장다리꽃무지
*Jumnos ferreroiminettiique*
p. 80

♀ 41mm
훼레로이미네티큐앞장다리꽃무지
*Jumnos ferreroiminettiique*
p. 80

♂ 52mm
룩케리노랑네점박이앞장다리꽃무지
*Jumnos ruckeri*
p. 81

# Identification Key

♂ 30mm
사슴풍뎅이
*Dicarnocephalus adamsi*
p. 67

♂ 31mm
부르케이앞장다리꽃무지
*Cheirolasia burkei burkei*
p. 68

♀ 27mm
부르케이앞장다리꽃무지
*Cheirolasia burkei burkei*
p. 68

♂ 31mm
부르케이셉텐트리오니스
앞장다리꽃무지
*Cheirolasia burkei septentrionis*
p. 69

♀ 28mm
부르케이셉텐트리오니스
앞장다리꽃무지
*Cheirolasia burkei septentrionis*
p. 69

♂ 48mm
데르비아나왕꽃무지
*Dicronorrhina derbyana derbyana*
p. 70

♀ 44mm
데르비아나왕꽃무지
*Dicronorrhina derbyana derbyana*
p. 70

♀ 30mm
그랄리뿔꽃무지
*Eudicella gralli gralli*
p. 74

♂ 46mm
그랄리움부로비타타뿔꽃무지
*Eudicella gralli umbrovittata*
p. 75

♀ 33mm
그랄리움부로비타타뿔꽃무지
*Eudicella gralli umbrovittata*
p. 75

♂ 40mm
스미씨쉬라티카뿔꽃무지
*Eudicella smithi shiratica*
p. 76

♀ 31mm
스미씨쉬라티카뿔꽃무지
*Eudicella smithi shiratica*
p. 76

♂ 39mm
트릴리네아뿔꽃무지
*Eudicella trilineata*
p. 77

♀ 33mm
트릴리네아뿔꽃무지
*Eudicella trilineata*
p. 77

♀ 46mm
룩케리노랑네점박이앞장다리꽃무지
*Jumnos ruckeri*
p. 81

♂ 60mm
크라치지줄무늬귀신꽃무지
*Mecynorhina kraatzi*
p. 82

♀ 49mm
크라치지줄무늬귀신꽃무지
*Mecynorhina kraatzi*
p. 82

♂ 60mm
사바게이점박이귀신꽃무지
*Mecynorhina savagei*
p. 83

♀ 52mm
사바게이점박이귀신꽃무지
*Mecynorhina savagei*
p. 83

# 모양찾기

♂ 68mm
주홍대왕귀신꽃무지 (유니칼라 형)
*Mecynorhina oberthueri* –Form *unicolor*
p. 84

♀ 54mm
주홍대왕귀신꽃무지 (유니칼라 형)
*Mecynorhina oberthueri* –Form *unicolor*
p. 84

♂ 60mm
주홍점박이대왕귀신꽃무지 (데코라타 형)
*Mecynorhina oberthueri* – Form *decorata*
p. 85

♀ 53mm
주홍점박이대왕귀신꽃무지 (데코라타 형)
*Mecynorhina oberthueri* – Form *decorata*
p. 85

♂ 69mm
토르콰타-우간덴시쓰대왕귀신꽃무지
*Mecynorhina torquata ugandensis*
p. 88

♀ 58mm
토르콰타-우간덴시쓰대왕귀신꽃무지
*Mecynorhina torquata ugandensis*
p. 88

♂ 64mm
토르콰타-우간덴시쓰대왕귀신꽃무지
*Mecynorhina torquata ugandensis*
p. 89

♀ 61mm
토르콰타-우간덴시쓰대왕귀신꽃무지
*Mecynorhina torquata ugandensis*
p. 89

♂ 40mm
하리시큰뿔꽃무지
*Megalorhina harrisi harrisihi*
p. 92

♀ 32mm
하리시큰뿔꽃무지
*Megalorhina harrisi harrisihi*
p. 92

♂ 45mm
하리시-엑스미아큰뿔꽃무지
*Megalorhina harrisi eximia*
p. 93

♀ 38mm
하리시-엑스미아큰뿔꽃무지
*Megalorhina harrisi eximia*
p. 93

♂ 19.5mm
로우예리쌍뿔꽃무지
*Mystroceros rouyeri*
p. 94

♀ 20mm
로우예리쌍뿔꽃무지
*Mystroceros rouyeri*
p. 94

# Identification Key

♂ 77mm
토르콰타-이마쿠리콜리스대왕귀신꽃무지
*Mecynorhina torquata immaculicollis*
p. 86

♀ 52mm
토르콰타-이마쿠리콜리스대왕귀신꽃무지
*Mecynorhina torquata immaculicollis*
p. 86

♂ 86mm
토르콰타-포게이대왕귀신꽃무지
*Mecynorhina torquata poggei*
p. 87

♀ 58mm
토르콰타-포게이대왕귀신꽃무지
*Mecynorhina torquata poggei*
p. 87

♂ 76mm
토르콰타-우간덴시스대왕귀신꽃무지
*Mecynorhina torquata ugandensis*
p. 90

♀ 56mm
토르콰타-우간덴시스대왕귀신꽃무지
*Mecynorhina torquata ugandensis*
p. 90

♂ 76mm
토르콰타-우간덴시스대왕귀신꽃무지
*Mecynorhina torquata ugandensis*
p. 91

♀ 56mm
토르콰타-우간덴시스대왕귀신꽃무지
*Mecynorhina torquata ugandensis*
p. 91

♂ 25mm
아우조욱시뿔꽃무지
*Neophaedimus auzouxi*
p. 95

♂ 36mm
스탄레이닙튠꽃무지
*Neptunides stanleyi*
p. 96

♀ 31mm
스탄레이닙튠꽃무지
*Neptunides stanleyi*
p. 96

♂ 23mm
폴리크로우스닙튠꽃무지
*Neptunides polychrous polychrous*
p. 97

♀ 23mm
폴리크로우스닙튠꽃무지
*Neptunides polychrous polychrous*
p. 97

♂ 31mm
폴리크로우스닙튠꽃무지
*Neptunides polychrous*
p. 98

♀ 31mm
폴리크로우스닙튠꽃무지
*Neptunides polychrous*
p. 98

# 모양찾기

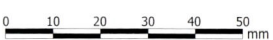

♂ 31mm
폴리크로우스넵튠꽃무지
*Neptunides polychrous*
p. 99

♀ 29mm
폴리크로우스넵튠꽃무지
*Neptunides polychrous*
p. 99

♂ 32mm
베르토로니흰큰머리꽃무지
*Rhamphorrhina bertolonii*
p. 100

♀ 27mm
베르토로니흰큰머리꽃무지
*Rhamphorrhina bertolonii*
p. 100

♂ 33mm
스프렌덴스흰큰머리꽃무지
*Rhamphorrhina splendens*
p. 101

♀ 22mm
스프렌덴스흰큰머리꽃무지
*Rhamphorrhina splendens*
p. 101

♀ 25mm
풍이
*Pseudotorynorrhina japonica*
p. 102

♂ 31mm
롱기셉스앞뿔꽃무지
*Taurrhina longiceps*
p. 106

♂ 31mm
롱기셉스앞뿔꽃무지
*Taurrhina longiceps*
p. 106

♂ 19mm
셉타노랑띠꽃무지
*Pedinorrhina septa*
p. 107

♀ 18mm
셉타노랑띠꽃무지
*Pedinorrhina septa*
p. 107

♂ 29mm
알페스트리스꽃무지
*Tmesorrhina alpestris*
p. 108

♂ 29mm
알페스트리스꽃무지
*Tmesorrhina alpestris*
p. 108

♂ 29.2mm
알페스트리-바후텐시스꽃무지
*Tmesorrhina alpestris bafutensis*
p. 109

♂ 33mm
후라메아풍이
*Torynorrhina flammea flammea*
p. 113

♂ 33mm
후라메아풍이
*Torynorrhina flammea flammea*
p. 113

♂ 34mm
후라메아풍이
*Torynorrhina flammea flammea*
p. 114

♀ 34mm
후라메아풍이
*Torynorrhina flammea flammea*
p. 114

♂ 29mm
후라메아풍이
*Torynorrhina flammea flammea*
p. 115

♂ 29mm
후라메아풍이
*Torynorrhina flammea flammea*
p. 115

♂ 26mm
에피피아타-환코이시주홍테꽃무지
*Pachnoda ephippiata francoisi*
p. 122

♂ 26mm
에피피아타-환코이시주홍테꽃무지
*Pachnoda ephippiata francoisi*
p. 122

♂ 26mm
마르기나타주홍테꽃무지
*Pachnoda marginata*
p. 123

♂ 26mm
마르기나타주홍테꽃무지
*Pachnoda marginata*
p. 123

♀ 22mm
쿠프레아점박이꽃무지
*Protaetia cuprea*
p. 124

♀ 22mm
쿠프레아점박이꽃무지
*Protaetia cuprea*
p. 124

# Identification Key

| ♀ 24mm | ♀ 37mm | ♂ 29mm | ♀ 27mm | ♂ 27mm | ♀ 27mm |
| --- | --- | --- | --- | --- | --- |
| 풍이 | 레스프렌덴스금광풍뎅이 | 구타타기린뿔꽃무지 | 구타타기린뿔꽃무지 | 줄리아기린뿔꽃무지 | 줄리아기린뿔꽃무지 |
| *Pseudotorynorrhina japonica* | *Rhomborhina resplendens* | *Stephanorrhina guttata* | *Stephanorrhina guttata* | *Stephanorrhina julia* | *Stephanorrhina julia* |
| p. 102 | p. 103 | p. 104 | p. 104 | p. 105 | p. 105 |

| ♂ 29.2mm | ♂ 29mm | ♂ 29mm | ♂ 22mm | ♂ 22mm | ♂ 37mm | ♂ 34mm |
| --- | --- | --- | --- | --- | --- | --- |
| 알페스트리-바후텐시스꽃무지 | 이리스긴몸광꽃무지 | 이리스긴몸광꽃무지 | 라에타연보석꽃무지 | 라에타연보석꽃무지 | 로칠디주걱턱꽃무지 | 로칠디주걱턱꽃무지 |
| *Tmesorrhina alpestris bafutensis* | *Tmesorrhina iris* | *Tmesorrhina iris* | *Tmesorrhina laeta* | *Tmesorrhina laeta* | *Trigonophorus rothschildi* | *Trigonophorus rothschildi* |
| p. 109 | p. 110 | p. 110 | p. 111 | p. 111 | p. 112 | p. 112 |

| ♂ 35mm | ♂ 35mm | ♂ 33mm | ♂ 33mm | 18mm | 17.5mm | ♂ 29mm | ♂ 29mm |
| --- | --- | --- | --- | --- | --- | --- | --- |
| 후라메아-키케리풍이 | 후라메아-키케리풍이 | 후라메아-키케리풍이 | 후라메아-키케리풍이 | 네점박이붉은띠꽃무지 | 네점박이붉은띠꽃무지 | 에피피아타-활케이주홍테꽃무지 | 에피피아타-활케이주홍테꽃무지 |
| *Torynorrhina flammea chicheryi* | *Torynorrhina flammea chicheryi* | *Torynorrhina flammea chicheryi* | *Torynorrhina flammea chicheryi* | *Glycyphana sp.* | *Glycyphana sp.* | *Pachnoda ephippiata falkei* | *Pachnoda ephippiata falkei* |
| p. 116 | p. 116 | p. 117 | p. 117 | p. 120 | p. 120 | p. 121 | p. 1121 |

| ♀ 20mm | ♀ 24mm | ♀ 23mm | ♂ 27mm | ♀ 28mm | ♂ 28mm | ♂ 28mm |
| --- | --- | --- | --- | --- | --- | --- |
| 쿠프레아-올리바세아꽃무지 | 쎄레비카점박이꽃무지 | 쎄레비카점박이꽃무지 | 녹스점박이큰꽃무지 | 녹스점박이큰꽃무지 | 루마위기점박이꽃무지 | 루마위기점박이꽃무지 |
| *Proteatia cuprea olivacea* | *Protaetia celebica* | *Protaetia celebica* | *Protaetia nox* | *Protaetia nox* | *Protaetia lumawigi* | *Protaetia lumawigi* |
| p. 125 | p. 126 | p. 126 | p. 127 | p. 127 | p. 128 | p. 128 |

# 모양찾기

♀ 21mm
헝가리카꽃무지
*Protaetia hungarica*
p. 129

♀ 21mm
헝가리카꽃무지
*Protaetia hungarica*
p. 129

♀ 24mm
괌꽃무지
*Protaetia guam*
p. 130

♀ 24mm
괌꽃무지
*Protaetia guam*
p. 130

♀ 21mm
오리엔탈리스점박이꽃무지
*Protaetia orientalis*
p. 131

♀ 20mm
오리엔탈리스점박이꽃무지
*Protaetia orientalis*
p. 131

♂ 23mm
흰점박이꽃무지
*Protaetia brevitarsis seulensis*
p. 132

♂ 27mm
우후리기큰점박이꽃무지
*Protaetia uhligi*
p. 136

♂ 27mm
우후리기큰점박이꽃무지
*Protaetia uhligi*
p. 136

♂ 23mm
베네라빌리스큰풀색꽃무지
*Protaetia venerabilis*
p. 137

♀ 22mm
베네라빌리스큰풀색꽃무지
*Protaetia venerabilis*
p. 137

♂ 26mm
샤우미꽃무지
*Sternoplus schaumii*
p. 138

♀ 26mm
샤우미꽃무지
*Sternoplus schaumii*
p. 138

♀ 26mm
샤우미꽃무지
*Sternoplus schaumii*
p. 139

♂ 28mm
스쿠아모수스뾰족투구꽃무지
*Mycteristes squamosus*
p. 148

♀ 23mm
스쿠아모수스뾰족투구꽃무지
*Mycteristes squamosus*
p. 148

♂ 28mm
볼렌호베니뾰족투구꽃무지
*Mycteristes vollenhoveni*
p. 149

♀ 24mm
볼렌호베니뾰족투구꽃무지
*Mycteristes vollenhoveni*
p. 149

♂ 25mm
하우드니투구꽃무지
*Phaedimus howdeni*
p. 150

♀ 21mm
하우드니투구꽃무지
*Phaedimus howdeni*
p. 150

♂ 22mm
하우드니투구꽃무지
*Phaedimus howdeni*
p. 151

♂ 16mm
아펠레스주홍줄홀쭉꽃무지
*Ixorida venerea apelles*
p. 158

♂ 15.5mm
아펠레스주홍줄홀쭉꽃무지
*Ixorida venerea apelles*
p. 158

♂ 14mm
솔로모니카주홍줄홀쭉꽃무지
*Ixorida solomonica*
p. 159

♀ 14mm
솔로모니카주홍줄홀쭉꽃무지
*Ixorida solomonica*
p. 159

♂ 19.2mm
프로핀쿠아흰점홀쭉꽃무지
*Ixorida propinqua*
p. 160

♀ 18.2mm
프로핀쿠아흰점홀쭉꽃무지
*Ixorida propinqua*
p. 160

# Identification Key

♂ 23mm
흰점박이꽃무지
*Protaetia brevitarsis seulensis*
p. 132

♂ 22mm
점박이꽃무지
*Protaetia orientalis submarmorea*
p. 133

♂ 22mm
점박이꽃무지
*Protaetia orientalis submarmorea*
p. 133

♂ 21mm
필리핀점박이꽃무지
*Protaetia phillppensis*
p. 134

♂ 21mm
필리핀점박이꽃무지
*Protaetia phillppensis*
p. 134

♀ 22mm
스켑시아점박이꽃무지
*Protaetia scepsia*
p. 135

♀ 20mm
스켑시아점박이꽃무지
*Protaetia scepsia*
p. 135

♂ 26mm
샤우미꽃무지
*Sternoplus schaumii*
p. 139

♀ 20mm
남미점박이모가슴꽃무지
*Gymnetis pantherina*
p. 142

♂ 21mm
지카니남미점박이모가슴꽃무지
*Gymnetis pantherina zikani*
p. 143

♀ 20mm
지카니남미점박이모가슴꽃무지
*Gymnetis pantherina zikani*
p. 143

♂ 19.2mm
르히노필루스투구꽃무지
*Mycteristes rhinophyllus*
p. 146

♂ 21mm
티베타나두뿔투구꽃무지
*Mycteristes tibetana*
p. 147

♂ 19mm
티베타나두뿔투구꽃무지
*Mycteristes tibetana*
p. 147

♂ 24mm
하우드니투구꽃무지
*Phaedimus howdeni*
p. 151

♂ 17mm
얼룩홀꽃무지
*Euselates sp.*
p. 154

♀ 19mm
스틱티카홀꽃무지
*Euselates stictica*
p. 155

♂ 22mm
엘레강스점박이홀쭉꽃무지
*Ixorida elegans*
p. 156

♂ 19mm
후리데리씨금박무늬홀쭉꽃무지
*Ixorida friderici*
p. 157

♂ 18.5mm
후리데리씨금박무늬홀쭉꽃무지
*Ixorida friderici*
p. 157

♂ 29mm
앤드로애디홀꽃무지
*Plectrone endroedii*
p. 161

♀ 28mm
앤드로애디홀꽃무지
*Plectrone endroedii*
p. 161

♂ 28mm
루마위기검은박쥐꽃무지
*Plectrone lumawigi*
p. 162

♀ 29mm
루마위기검은박쥐꽃무지
*Plectrone lumawigi*
p. 162

♂ 23mm
트리칼라홀쭉꽃무지
*Taeniodera tricolor tricolor*
p. 163

♀ 20mm
트리칼라홀쭉꽃무지
*Taeniodera tricolor tricolor*
p. 163

# 꽃무지아과   Subfamily Cetoniinae

꽃무지아과는 10아족으로 나뉘어지나 세 개 족(Cremastocheilini족, Xiphoscelidini족, Diplongnathini족)을 제외 한 나머지 일곱 개 족을 수록하였다.

This family is divided into 10 subgenera. Contained in this book are 7, excluding Cremastocheilini, Xiphoscelidini and Diplongnathini.

## 1. 개미집살이꽃무지족(族) Cremastocheilini

이 족은 약 50속 390종이 알려져 있으며 유라시아서부, 오스트레일리아, 마다가스카르를 제외한 전 세계에 서식하고 있다. 주로 개미집 속에 들어가 개미의 유충을 먹고 살기 때문에 체형도 그에 맞게 발달되었다.

This tribe is known to be divided into 390 species under some 50 genera, breeding worldwide except the western part of Eurasia, Australia and Madagascar. Usually penetrating formicaries and feeding on larval ants, it has evolved its body shape in a way that facilitates such action.

## 2. 거짓개미집살이꽃무지족(族) Xiphoscelidini

이 족은 생태가 잘 알려져 있지 않은 족으로 16속 약 50종으로 알려져 있다. 마다가스카르와 오스트레일리아의 소수의 종을 제외하고는 대부분 아프리카에 서식하고 있다.

This tribe, about which many things have yet to be discovered, is divided into some 50 species under 16 genera. Most of the individuals in this tribe inhabit Africa while a small number lives in Madagascar and Australia.

## 3. 마다가스카르꽃무지족(族) Stenotarsiini

이 족은 56속 362종 이상, 알려져 있으며 아프리카 대륙에 2속 2종을 제외하고, 모두가 마다가스카르 섬에 서식하고 있다.
이 장에서는 1속(genus) 1종(species) 2개체를 수록하였다.

More than 362 species are known to exist under the 56 genera of this tribe. All of them but the 2 species under 2 genera living in Africa inhabit Madagascar. 2 individuals of 1 genus, 1 species are included in this section.

## 4. 오스트레일리아꽃무지족(族) Schizorhinini

이 족은 42속 약370종이 알려져 있으며 대부분이 오스트레일리아구(Australian region)에 서식하고 있다.
이 장에서는 1속(genus) 1종(species) 2개체를 수록하였다.

Approximately 370 species are known to exist under the 42 genera of this tribe. Most of them inhabit the Australian region.
2 individuals of 1 genus, 1 species are included in this section.

Section 1
# Cremastocheilini

Section 2
# Xiphoscelidini

Section 3
# Stenotarsiini

Section 4
# Schizorhinini

## 꽃무지아과 / Cetoniinae / Tribe 3. Stenotarsiini

**Euchroea**

| 학 명 Scientific name | 채집국 Collected locality | 크 기 Size | 참 고 Remark |
|---|---|---|---|
| 노랑마다가스카르꽃무지<br>*Euchroea auripimenta* (Gory & Percheron, 1835) | 마다가스카르<br>Madagascar | ♂♀ 25-32mm | |

마다가스카르 섬에 서식한다. 몸빛은 전체적으로 황색바탕이며 머리와 소순판에 한 개, 가슴과 미절판에 한 쌍, 딱지날개에 두 쌍의 검은 반점을 가지고 있다. 광택이 없다.

Inhabiting Madagascar, they have a yellow body without gloss. They have one speckle on each of the cephalic part and scutellum, a pair of speckles on each of the thorax and pygidium and two pairs of speckles on the elytra (hard wings).

[Distribution] Madagascar

● 채집지 Collected Locality
● 분포지 Distribution

마다가스카르
Madagascar I.

♂

마다가스카르 산 29mm
(Ambatolampy, Madagascar. 2003. 3.)

♀

마다가스카르 산 32mm
(Ambatolampy, Madagascar. 2003. 3.)

# Flower beetles / Tribe 4. Schizorhinini

## Trichaulax

| 학 명 Scientific name | 채집국 Collected locality | 크 기 Size | 참 고 Remark |
|---|---|---|---|
| 마크래이꽃무지<br>*Trichaulax macleayi* Kraatz, 1894 | 오스트레일리아<br>Australia | ♂♀ 26-36mm | |

오스트레일리아에 서식한다. 머리와 가슴은 검정색으로 광택이 있으며 딱지날개는 암적색 바탕에 황색 강모가 나있는 세 쌍의 세로줄이 있다.

They inhabit Australia. Their black cephalic part and thorax have no gloss while the dark red elytra (hard wings) have three pairs of yellow-bristled stripes.

[Distribution] Australia

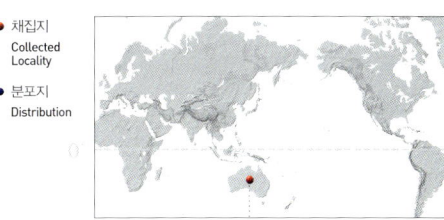

● 채집지 Collected Locality
● 분포지 Distribution

오스트레일리아 Australia

♀

오스트레일리아 산 27mm
(Mt.Garmet, NQ. Australia. 2004. 9.)

♂

오스트레일리아 산 29mm
(Mt.Garmet, NQ. Australia. 2004. 9.)

# 골리앗대왕꽃무지족(族)

이 족은 90속 410종이 알려져 있으며 꽃무지 최고의 대형종을 비롯하여 수컷들이 뿔을 가지는 특징을 가진다. 대부분 에티오피아구(Ethiopian region)와 동양 열대구(Oriental region)에 서식하며 마다가스카르와 멕시코에서는 9종만이 알려져 있다.
이 장에서는 25속(genus) 59종(species) 132개체를 수록하였다.

Currently, 410 species of 90 genera are known to be classified into this tribe. Including the biggest Cetoniid species, its males are characterized by their horns. The majority of them inhabit Ethiopian region and Oriental region while only 9 species have been discovered in Madagascar and Mexico.
132 individuals of 25 genus, 59 species are included in this section.

Section 5
Goliathini

## 꽃무지아과 / Cetoniinae

**Goliathus**

| 학 명 Scientific name | 채집국 Collected locality | 크 기 Size | 참 고 Remark |
|---|---|---|---|
| 골리앗레기우스대왕꽃무지<br>*Goliathus regius* (Klug, 1835) | 코트디부아르<br>Cote divoire | ♂ 58-115mm, ♀ 56-82mm | Subtribe Goliathina |

아프리카 서부에 넓게 서식한다. *G. goliatus*와 더불어 세계 최대의 무게를 가지고 있다. 살아 있을 때의 체중은 100g정도로 무겁다. 몸빛은 수컷은 광택이 없으며 암컷은 약한 광택을 지니고 있다. 머리와 가슴, 딱지날개 모두 백색바탕에 검정색의 무늬를 가진다. 가슴은 검정색을 띤 한 쌍의 반점과 두 쌍의 세로줄무늬를 가지고 있으나 가운데 한 쌍의 세로줄 무늬가 가장 크다. 넓적다리마디와 가운데가슴복판은 적갈색을 띠며 복마디는 암녹색으로 모두 광택을 지닌다.

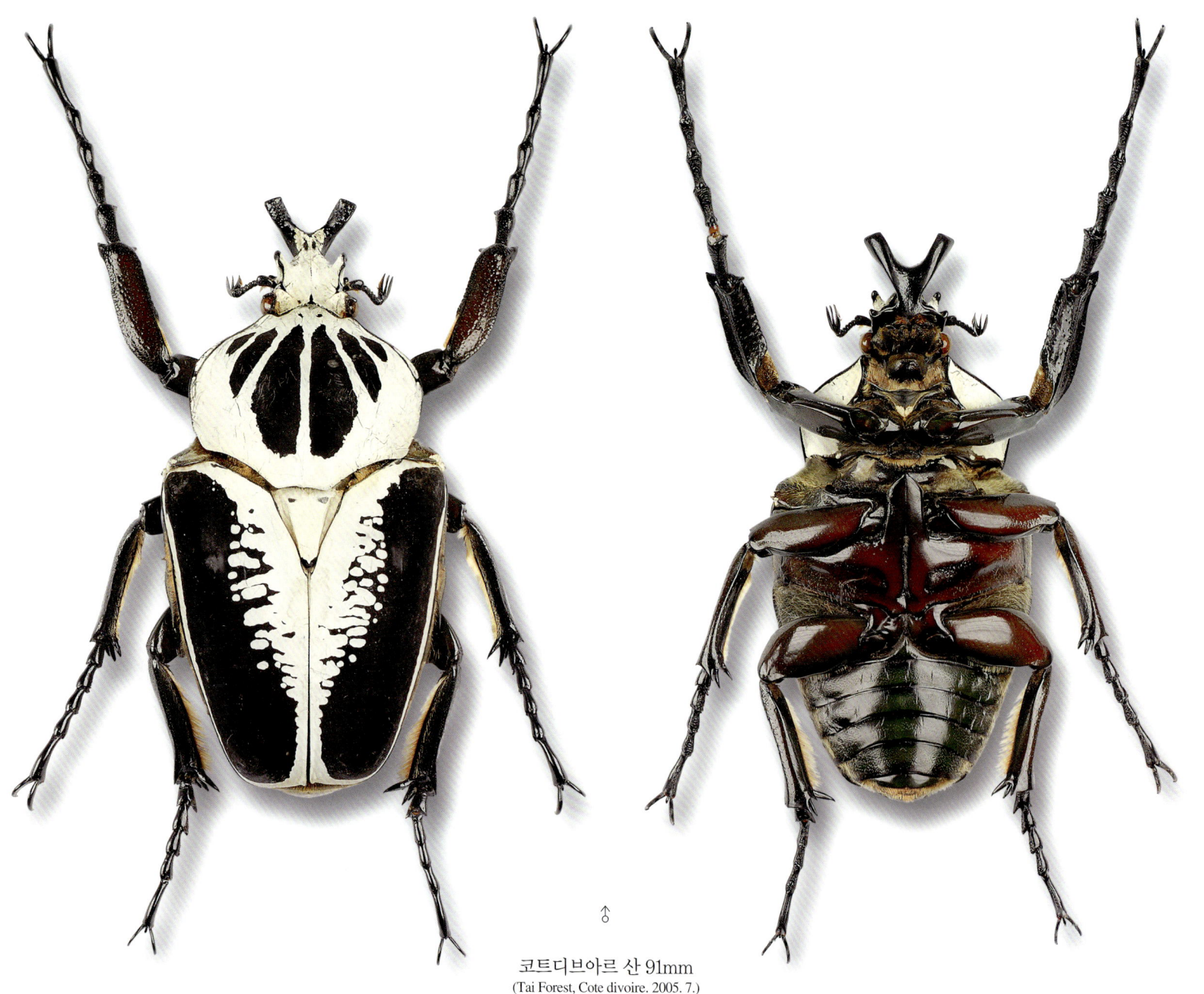

♂
코트디브아르 산 91mm
(Tai Forest, Cote divoire. 2005. 7.)

# Flower Beetles / Tribe 5. Goliathini

This species, which has a wide habitat across the western part of Africa, constitutes the two heaviest insect species of the world along with *G. goliatus*. It weighs about 100g when alive. The males have no gloss while the females are slightly glossy. The cephalic part, thorax and elytra (hard wings) are white with similar black patterns, while the thorax has a pair of black speckles and two pairs of stripes, of which the inner pair of stripes has the biggest pattern. The thigh joints and mesosternum are dark-green and have a metallic gloss.

[Distribution] Sierra Leone, Guinea, Ghana, Benin, Cote divoire, Togo, Nigeria

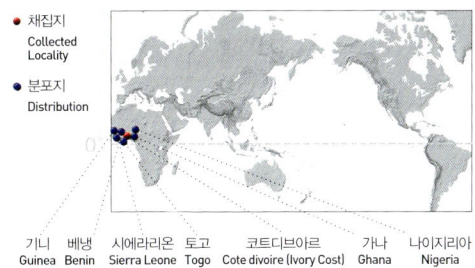

♀

코트디브아르 산 63mm
(Tai Forest, Cote divoire. 2005. 7.)

♂

코트디브아르 산 63mm
(Tai Forest, Cote divoire. 2005. 7.)

## 꽃무지아과 / Cetoniinae

**Goliathus**

| 학 명 Scientific name | 채집국 Collected locality | 크 기 Size | 참 고 Remark |
|---|---|---|---|
| 골리앗대왕꽃무지<br>*Goliathus Goliathus* (Drury, 1770) | 콩고민주공화국<br>D.R. Congo | ♂ 55-110mm, ♀ 54-80mm | Subtribe Goliathina |

중앙아프리카 중서부에 서식한다. *Goliathus*속은 *G. goliathus, G. regius, G. orientalis, G. cacicus, G. albosignatus* 다섯 종이 있다. 이 종은 *G. regius*와 더불어 곤충 중에서 세계 최대의 몸무게를 자랑한다. 살아있을 때의 체중은 100g정도로 무겁다. 서식지는 기온이 35℃, 습도 60~70%의 고온 다습한 환경으로 표고 1,300m 전후에서 발견된다. 몸빛은 수컷은 광택이 없으며 암컷은 약한 광택을 지니고 있다. 머리와 가슴은 백색바탕에 적갈색을 띤 한 쌍의 반점과 두 쌍의 넓은 세로줄무늬를 가지고 있다. 딱지날개는 적갈색을 띠며 백색의 띠와 같이 무늬와 반점이 둘러져 있으나 이어져 있지는 않다. 복면의 넓적다리마디와 가운데가슴복판은 녹색으로 금속성 광택을 지닌다. 변이가 다양하여 goliathus, apicalis, conspersus, undulus, albatus, quadrimaculatus, vittatus 일곱가지 형이 알려져 있다.

콩고민주공화국 산 102mm
(N.Kive, Lake, D.R.Congo. 2003. 3.)

# Flower Beetles / Tribe 5. Goliathini

The *Goliathus* genus is divided into five species.(*G. goliathus*, *G. regius*, *G. orientalis*, *G. cacicus*, *G. albosignatus*) This species, which inhabits the mid-western part of Central Africa, is the heaviest insect species of the world, comparable to *G. regius*. Weighing about 100g while alive, it lives in a hot and humid habitat with an average temperature of 35°C and average humidity of 60% to 70%. It is usually discovered at an altitude of 1,300m or higher. The males have no gloss while the females are slightly glossy. On their white cephalic part and thorax are a pair of reddish-brown speckles and two pairs of wide stripes. The reddish-brown elytra (hard wings) are patterned and speckled like a disconnected white strand. Their thigh joints and mesosternum are green and have a metallic gloss. This species is divided into seven form.(goliathus, apicalis, conspersus, undulus, albatus, quadrimaculatus, vittatus)

[Distribution] Nigeria, Cameroon, Central Africa Rep., Gabon, D.R. Congo, Uganda, Western Kenya, Tanzania

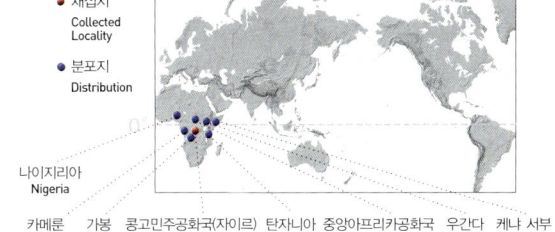

♀

콩고민주공화국 산 76mm
(N.Kive, Lake, D.R.Congo. 2003. 7.)

♂

콩고민주공화국 산 65mm
(N.Kive, Lake, D.R.Congo. 2004. 7.)

## 꽃무지아과 / Cetoniinae

### Goliathus

| 학 명 Scientific name | 채집국 Collected locality | 크 기 Size | 참 고 Remark |
|---|---|---|---|
| 골리앗대왕꽃무지 (아피카리스 형)<br>*Goliathus goliathus* (Drury, 1770) - Form apicalis | 콩고민주공화국<br>D.R. Congo | ♂ 55-110mm, ♀ 54-80mm | Subtribe Goliathina |

중앙아프리카에 서식한다. 이 종은 *G. goliatus*의 아피카리스 형이다. *G. goliatus*와 비교하여 딱지날개 양쪽 테두리에 백색의 얇은 띠가 더 길고 반점이 많아 구분된다. 몸빛은 수컷은 광택이 없으며 암컷은 약한 광택을 지니고 있다. 머리와 가슴은 백색바탕에 적갈색을 띤 한 쌍의 반점과 두 쌍의 넓은 세로줄무늬를 가지고 있다. 딱지날개는 적갈색을 띠며 백색의 띠가 둘러져 있으나 이어져 있지는 않다. 복면의 넓적다리마디와 가운데가슴복판은 녹색과 함께 갈색으로 금속성 광택을 지닌다.

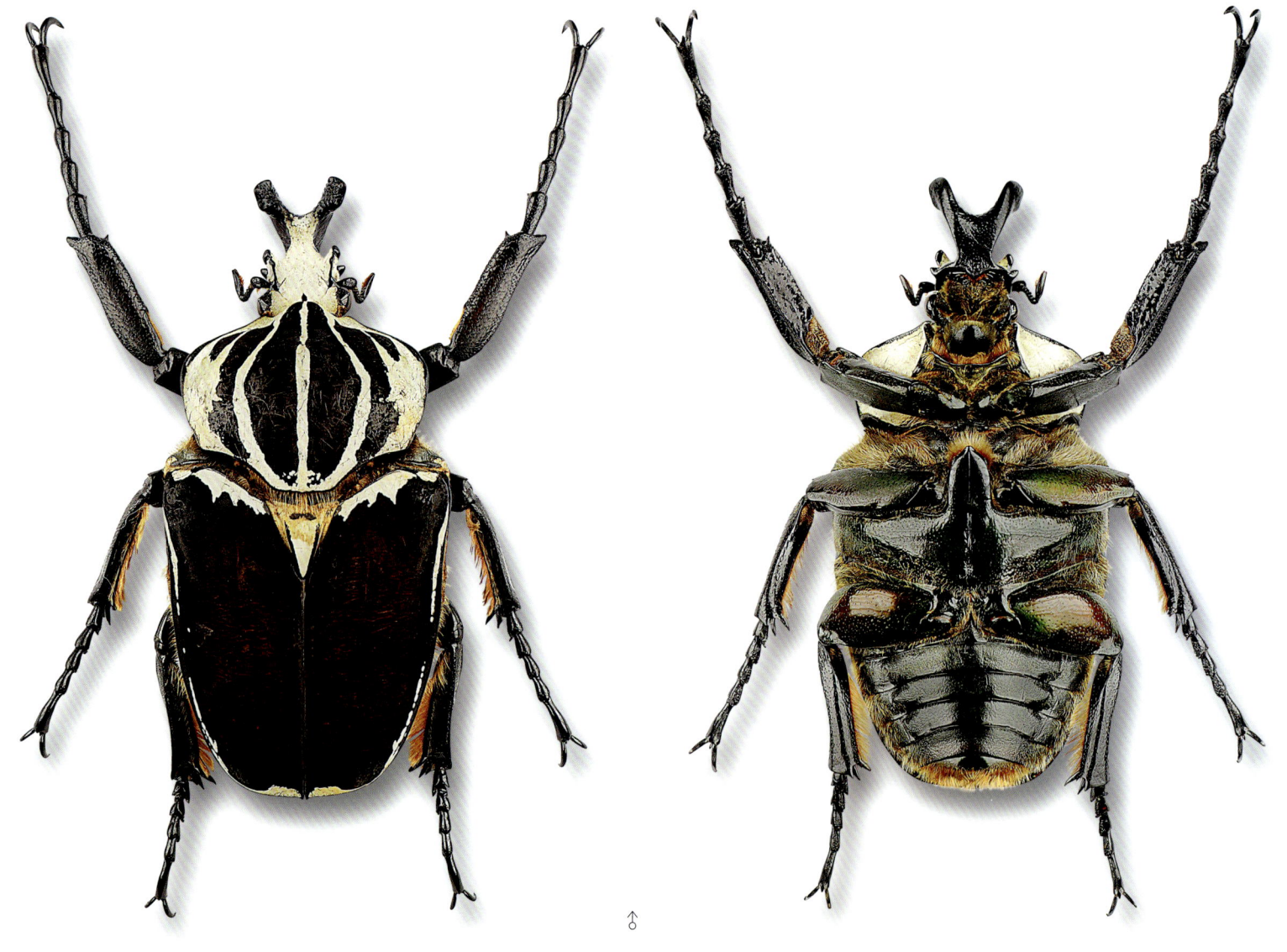

♂

콩고민주공화국 산 97mm
(N.Kive, Lake, D.R.Congo. 2004. 3.)

# Flower Beetles / Tribe 5. Goliathini

This species, which inhabits Central Africa, constitutes the form of *G. goliatus* with apicalis. Compared with those of *G. goliatus*, its elytra (hard wings) have disconnected edges on their both sides while the latter have more white stripes and speckles, facilitating the distinction between the two subspecies. Its males are not glossy while its females are slightly glossy. Its white cephalic part and thorax have a pair of red-brown speckles and two pairs of wide stripes. The elytra (hard wings) are also red-brown with some white stripes surrounding them but disconnected. Its thigh joints and mesosternum are green and brown with a metallic gloss.

[Distribution] D.R. Congo

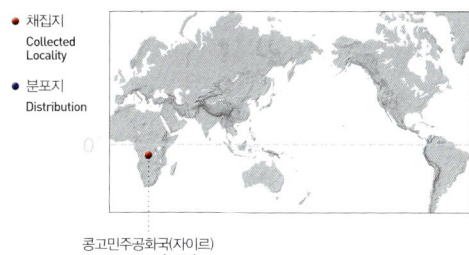

D.R. Congo(Zaire)

♀

콩고민주공화국 산 77mm
(N.Kive, Lake, D.R.Congo. 2003. 3.)

♂

콩고민주공화국 산 88mm
(N.Kive, Lake, D.R.Congo. 2004. 3.)

## 꽃무지아과 / Cetoniinae

**Goliathus**

| 학 명 Scientific name | 채집국 Collected locality | 크 기 Size |
|---|---|---|
| 골리앗대왕꽃무지 (콘스페르서스 형)<br>*Goliathus goliathus* (Drury, 1770) **- Form conspersus** |  콩고민주공화국 D.R. Congo    콩고 Congo | ♂ 54-110mm, ♀ 54-80mm |

중앙아프리카에 서식한다. *G. goliatus*의 콘스페르서스 형이다. *G. goliatus*와 비교하여 오히려 *G. orientalis*에 가까운 색깔과 무늬를 가지고 있다. 몸빛은 수컷은 광택이 없으며 암컷은 약한 광택을 지니고 있다. 가슴은 검정색을 띤 한 쌍의 반점과 두 쌍의 넓은 세로줄무늬를 가지고 있으나 두번째 세로줄무늬는 완벽하지 않다. 딱지날개는 검정색의 조밀한 망사무늬를 가지고 있다. 복면의 넓적다리마디와 가운데가슴복판은 갈색으로 금속성 광택을 지닌다.

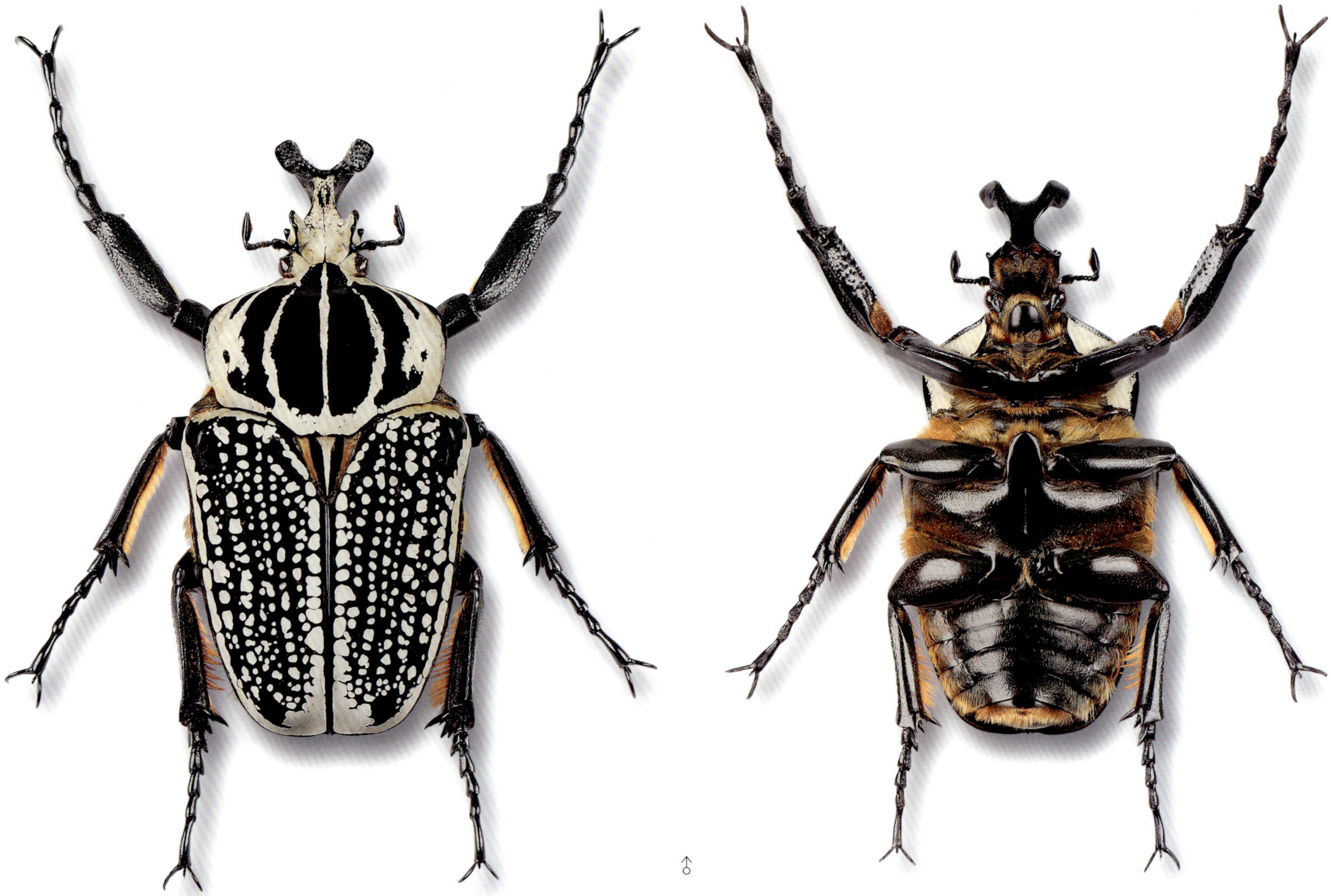

콩고민주공화국 산 94mm
(N.Kive, Lake, D.R. Congo. 2004. 7.)

## Flower Beetles / Tribe 5. Goliathini

참 고 Remark

Subtribe Goliathina

This species, which inhabits Central Africa, is a form of *G. goliatus* with conspersus. When compared with *G. goliatus*, this species has colors and patterns more similar to those of *G. orientalis*. The males have no gloss while the females are slightly glossy. Though a pair of speckles and two pairs of wide stripes are seen on the thorax, the second pair of stripes is not clear. On the elytra (hard wings) are some dense and black mesh patterns. Their femur and mesosternum are brown and have a metallic gloss.

[Distribution]  D.R. Congo, Congo

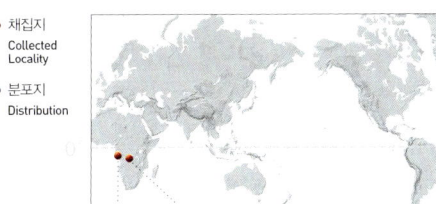

콩고　　　콩고민주공화국(자이르)
Congo　　D.R. Congo(Zaire)

♀

콩고민주공화국 산 73mm
(N.Kive, Lake, D.R.Congo. 2004. 7.)

♀

콩고 산 75mm
(Likas, Shaba, SE. Congo. 2004. 3.)

## 꽃무지아과 / Cetoniinae

**Goliathus**

| 학 명 Scientific name | 채집국 Collected locality | 크 기 Size | 참 고 Remark |
|---|---|---|---|
| 골리앗대왕꽃무지 (비타투스 형)<br>*Goliathus goliathus* (Voet, 1779) - **Form vittatus** | 콩고민주공화국<br>D.R. Congo | ♀ 45mm | Subtribe Goliathina |

중앙아프리카 서부에 서식한다. 이 종은 *G. goliatus*의 비타투스 형이다.
This species form of *G. goliatus* with vittatus. inhabits the western part of Central Africa.

[Distribution]   D.R. Congo

● 채집지 Collected Locality
● 분포지 Distribution

콩고민주공화국(자이르)
D.R. Congo (Zaire)

♀

콩고민주공화국 산 45mm
(N.Kivu Lake, D.R. Congo. 2002.12.)

Flower Beetles / Tribe 5. Goliathini

## Goliathus

| 학 명 Scientific name | 채집국 Collected locality | 크 기 Size | 참 고 Remark |
|---|---|---|---|
| 골리앗대왕흰꽃무지 (푸스투라투스 형)<br>*Goliathus orientalis* Moser, 1909 - Form pustulatus | 콩고<br>Congo | ♂ 55-108mm, ♀ 60-85mm | Subtribe Goliathina |

중앙아프리카 서부에 서식하며, 이 종은 orientalis, preiss, pustulatus, undulatus 네가지 형이 있다. 딱지날개의 무늬는 *G. orientalis*와 비교하여 조밀한 망사무늬가 아닌 갈라진 땅바닥과 같은 무늬를 가지고 있다.

This species inhabits the western part of Central Africa. Its elytra (hard wings) pattern looks like cracked ground, different from that of *G. orientalis*, which is mesh-like and black. This species is divided into fore form.(orientalis, preiss, pustulatus, undulatus)

[Distribution] Congo

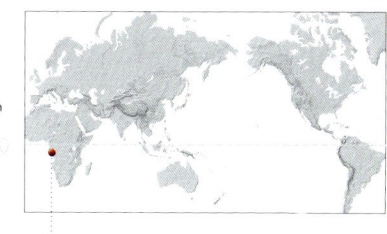

콩고
Congo

♂
콩고 산 78mm
(Congo. 2004.)

♂
콩고 산 82mm
(Congo. 2004.)

# 꽃무지아과 / Cetoniinae

## Goliathus

| 학 명 Scientific name | 채집국 Collected locality | 크 기 Size | 참 고 Remark |
|---|---|---|---|
| 골리앗오리엔탈흰대왕꽃무지<br>*Goliathus orientalis* Moser, 1909 |  콩고민주공화국<br>D.R. Congo | ♂ 55-108mm, ♀ 60-85mm | Subtribe Goliathina |

중앙아프리카에 서식한다. 몸빛은 수컷은 광택이 없으며 암컷은 약한 광택을 지니고 있다. 머리와 가슴, 딱지날개 모두 백색바탕에 검정색의 무늬를 가진다. 가슴은 검정색을 띤 한 쌍의 반점과 두 쌍의 세로줄무늬를 가지고 있다. 딱지날개는 검정색의 조밀한 망사무늬를 가지고 있다. *G. g. conspersus*와 비교하여 앞가슴복판 양쪽에 검정색의 반점을 가지고 있어 구분된다. 각각의 다리마디와 가운데 가슴복판, 복마디는 흑갈색으로 광택을 지닌다.

콩고민주공화국 산 90mm
(Likas, Shaba, D.R. Congo. 2004. 3.)

# Flower Beetles / Tribe 5. Goliathini

This species, which inhabits Central Africa, the males have no gloss while the females are slightly glossy. The cephalic part, thorax and elytra (hard wings) are white with black patterns all alike while the thorax has a pair of black speckles and two pairs of stripes. The elytra (hard wings) have a dense and black pattern. The black speckles on both sides of the thoracic sternum are what distinguish this species *G. goliathus* conspersus. Its leg joints, prosternum and abdominal sternites are glossy and brownish black.

[Distribution]  Western D.R. Congo, Western Tanzania, Northern Angola, Zambia

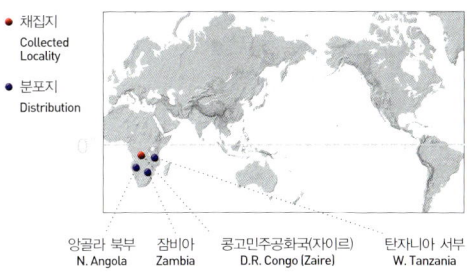

♀

콩고민주공화국 산 57mm
(Likas, Shaba, D.R. Congo. 2004. 3.)

♀

콩고민주공화국 산 70mm
(Likas, Shaba, D.R. Congo. 2004. 3.)

## 꽃무지아과 / Cetoniinae

**Goliathus**

| 학 명 Scientific name | 채집국 Collected locality | 크 기 Size | 참 고 Remark |
|---|---|---|---|
| 골리앗흰대왕꽃무지 (프레이시 형)<br>*Goliathus orientalis* Moser, 1909 - Form preissi | 탄자니아<br>Tanzania | ♂ 55-108mm, ♀ 60-85mm | Subtribe Goliathina |

중앙아프리카 중서부에 서식한다. 딱지날개의 무늬는 *G. orientalis*와 비교하여 조밀한 망사무늬가 아닌 세로줄무늬를 가진다. 몸빛은 수컷은 광택이 없으며 암컷은 약한 광택을 지니고 있다. 머리와 가슴, 딱지날개 모두 백색바탕에 검정색의 무늬를 가진다. 가슴은 검정색을 띤 한 쌍의 반점과 두 쌍의 세로줄무늬를 가지고 있다. 딱지날개는 검정색으로 세로줄 무늬를 가지고 있다. 가슴복판과 배 둘레에 나있는 털들은 *G. orientalis*와 비교하여 더 길고 조밀하며 밝은 황색을 띤다. 각각의 다리마디와 가운데가슴복판, 복마디는 갈색 광택을 지닌다.

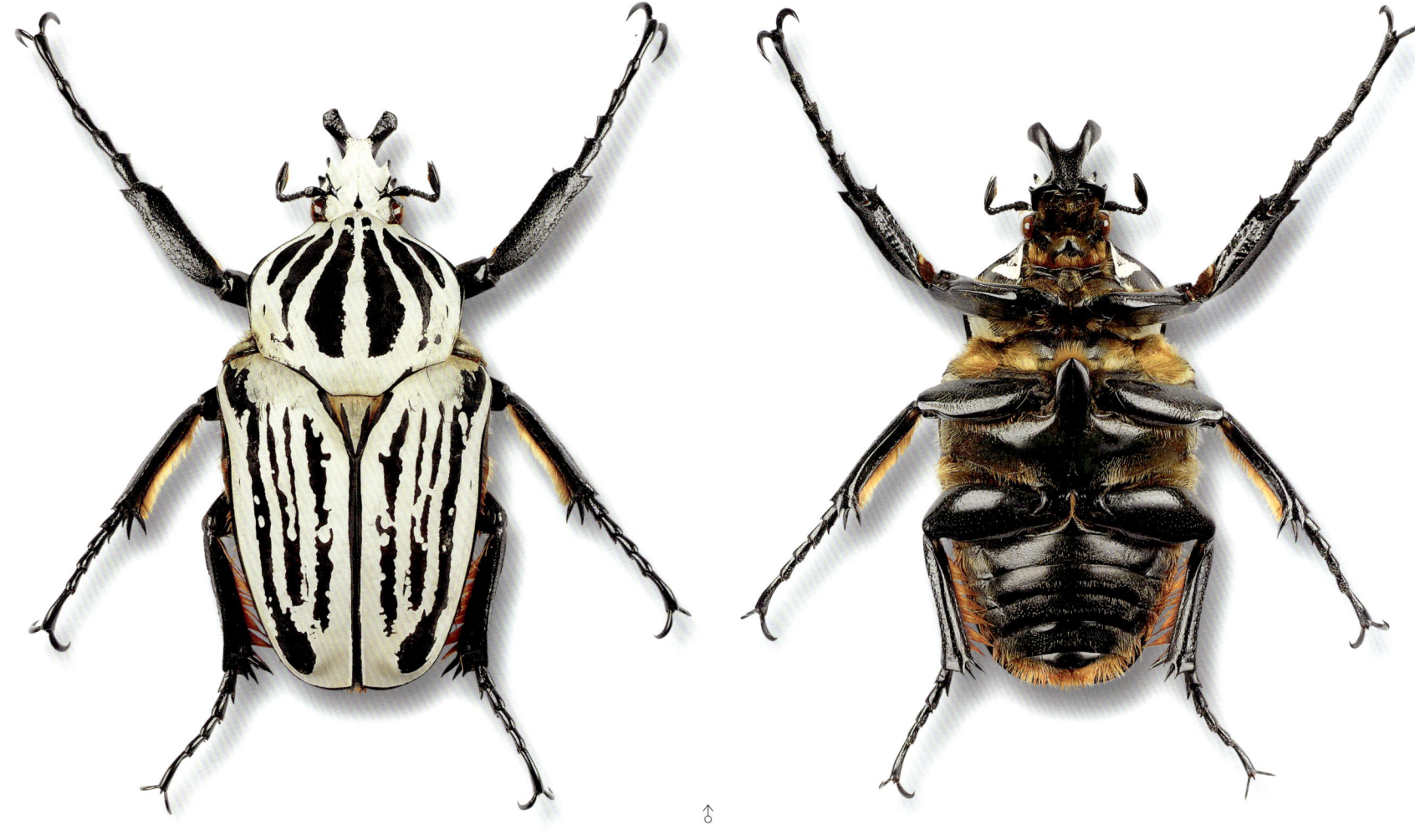

♂

탄자니아 산 75mm
(Amani Forest, Tanzania. 2006. 3.)

# Flower Beetles / Tribe 5. Goliathini

This species, which inhabits the mid-western part of Central Africa, the elytra (hard wings) have some stripes, unlike those of G. orientalis, which have a dense mesh-like pattern. The males have no gloss while the females are slightly glossy. The cephalic part, thorax and elytra (hard wings) are white with similar black patterns, while the thorax has a pair of black speckles and two pairs of stripes. The black elytra (hard wings) have stripes. The bright-yellow hair on the thoracic sternum and around the abdomen is longer and denser than that of *G. orientalis*. Their leg joints, mesosternum and abdominal sternites are glossy and brown.

[Distribution] Tanzania(Mts. Usambara)

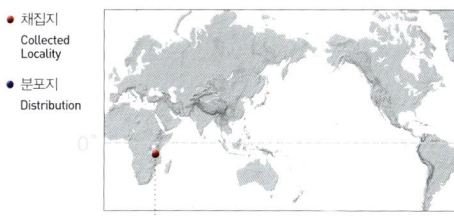

탄자니아(우삼바라 산악지역)
Tanzania(Mts. Usambara)

♀

탄자니아 산 74mm
(Mts. Usmbara Tanzania 2002. 5.)

♂

탄자니아 산 59mm
(Amani Forest, Tanzania. 2006. 3.)

## 꽃무지아과 / Cetoniinae

**Goliathus**

| 학 명 Scientific name | 채집국 Collected locality | 크 기 Size | 참 고 Remark |
|---|---|---|---|
| 골리앗흰대왕꽃무지 (운두라투스 형)<br>*Goliathus orientalis* Moser, 1909 - Form undulatus | 콩고민주공화국<br>D.R. Congo | ♂ 55-108mm, ♀ 60-85mm | Subtribe Goliathina |

중앙아프리카 중서부에 서식한다. 딱지날개의 무늬는 *G. orientalis*와 비교하여 조밀한 망사무늬가 아닌 보다 덜 조밀한 무늬를 지니며 앞가슴복판 양쪽에 검정색의 반점을 가지고 있지 않아 구분된다. 수컷의 몸빛은 광택이 없으며 암컷은 약한 광택을 지니고 있다. 머리와 가슴, 딱지날개 모두 백색바탕에 검정색의 무늬를 가진다. 가슴은 검정색을 띤 한 쌍의 반점과 두 쌍의 세로줄무늬를 가지고 있으나 두 번째의 것은 끊어져 있다. 각각의 다리마디와 가운데가슴복판, 복마디는 흑갈색 광택을 지닌다.

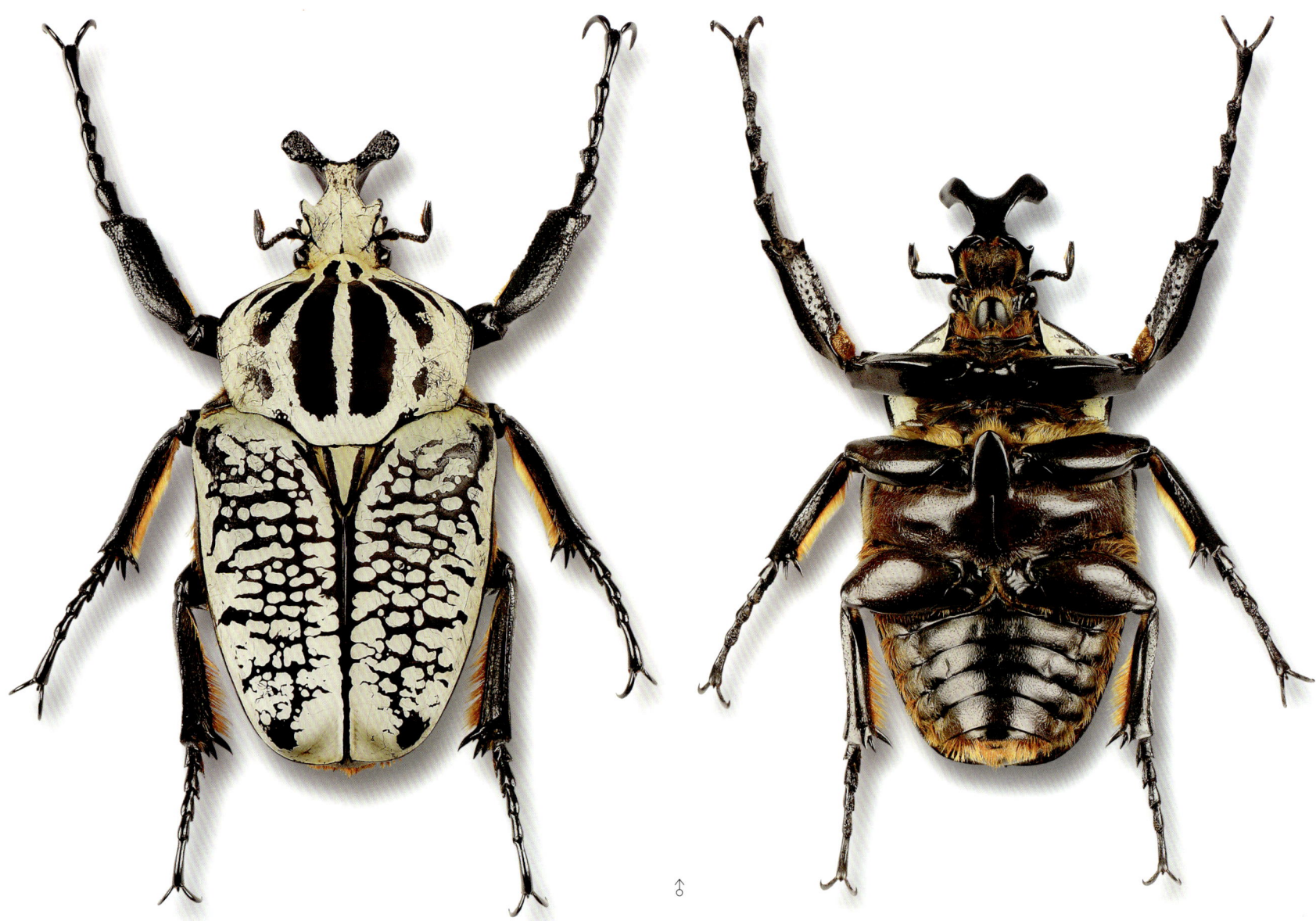

♂

콩고민주공화국 산 87mm
(Likas, Shaba, D.R. Congo. 2005. 5.)

# Flower Beetles / Tribe 5. Goliathini

This species, which inhabits the mid-western part of Central Africa, the elytra (hard wings) pattern is less dense than that of G. orientalis. The lack of black speckles on both sides of the prosternum also distinguishes it from G. orientalis. The males have no gloss while the females are slightly glossy. The cephalic part, thorax and elytra (hard wings) are white with black patterns all alike while the thorax has a pair of black speckles and two pairs of stripes, the second of which is disconnected. Their leg joints, mesosternum and abdominal sternites are glossy and brownish black.

[Distribution] Western D.R. Congo

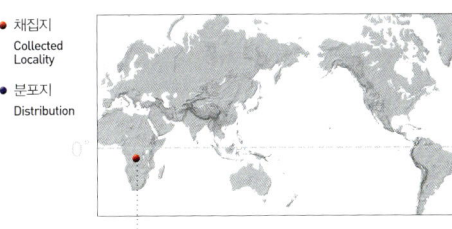

콩고민주공화국(자이르)
D.R. Congo (Zaire)

♀

콩고민주공화국 산 62mm
(Likas, Shaba, D.R. Congo. 2005. 5.)

♂

콩고민주공화국 산 68mm
(Likas, Shaba, D.R. Congo. 2005. 5.)

## 꽃무지아과 / Cetoniinae

| 학 명 Scientific name | 채집국 Collected locality | 크 기 Size |
|---|---|---|
| 골리앗알보씨그나투스-키르키아누스대왕꽃무지<br>*Goliathus albosignatus kirkianus* Gray,1864 |  탄자니아 Tanzania     짐바브웨 Zimbabwe | ♂ 43-71mm, ♀ 45-55mm |

남아프리카 북동부에 서식한다. 이 종은 *G. a. albosignatus*와 더불어 두 개의 아종이 있으며 *Goliathus*속 중 가장 작은 아종이다. 수컷의 머리뿔은 마치 V자처럼 발달하여 위쪽으로 굽어있다. 이 종의 몸빛은 광택이 없는 황백색을 띠며 검정색 또는 암갈색을 띤 호피무늬를 가지고 있다. 가운데종아리마디와 뒷다리 종아리마디에 나있는 강모가 황색인데 반하여, *G. a. albosignatus*는 흑색으로 이 두 아종의 구분하는데 결정적 요소가 된다.

탄자니아 산 66mm
(Tanzania. 2001. 11.)

탄자니아 산 65mm
(Tanzania. 2002. 11.)

# Flower Beetles / Tribe 5. Goliathini

### 참 고 Remark

Subtribe Goliathina

This species inhabits the northeastern part of South Africa. This species, the smallest of the *Goliathus* genus, has two subgenera together with *G. a. albosignatus*. The V-shaped cephalic horns of the males are bent upward. This species' body is yellowish white with no gloss. Its yellow body with a black or dark brown tiger-skin pattern is what separates it from *G. a. albosignatus*, another subspecies which has a black body.

[Distribution] Tanzania, Northern region of Zambian river valleys, Malawi, Northern Zimbabwe

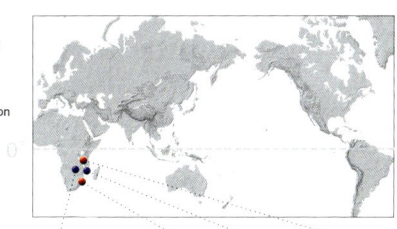

♀

짐바브웨 산 42mm
(Zimbabwe. 2004. 11.)

♂

짐바브웨 산 58mm
(Zimbabwe. 2004. 11.)

## 꽃무지아과 / Cetoniinae

**Goliathus**

| 학 명 Scientific name | 채집국 Collected locality | 크 기 Size | 참 고 Remark |
|---|---|---|---|
| 골리앗카시쿠스대왕꽃무지<br>*Goliathus cacicus* (Voet, 1779) | 코트디부아르<br>Cote divoire | ♂ 56-98mm, ♀ 58-79mm | Subtribe Goliathina |

아프리카 중서부에 서식한다. 이 종의 몸빛은 수컷의 경우 머리와 가슴은 황색바탕에 세 쌍의 검정색 세로줄 무늬를 가지고 있다. 딱지날개는 회색으로 상부와 하부 양쪽으로 각각 검정색의 반점을 가지고 있으며 상부는 크고, 하부의 것은 작다. 수컷의 머리뿔은 마치 V자 처럼 발달하여 위쪽으로 굽어있으며 개체의 크기에 따라 뿔의 크기도 달라진다. 각각의 다리와 복면은 암갈색을 띠며 광택을 지닌다. 암컷의 몸빛은 광택을 지니고 있다.

♂
코트디부아르 산 79mm
(Akupe, Cote divoire. 2005. 9.)

# Flower Beetles / Tribe 5. Goliathini

The males of this species, which inhabits the mid-western part of Africa, have a yellow cephalic part and thorax with three pairs of black stripes. Its grey elytra (hard wings) have black speckles on both the upper and lower sides, and those on the former are bigger than the ones on the latter. The males' V-shaped cephalic horns are bent upward and their size is dependent upon each individual's size. Their legs and back are dark brown and glossy. The females have a glossy body.

[Distribution]  Liberia, Guinea, Cote divoire, Ghana

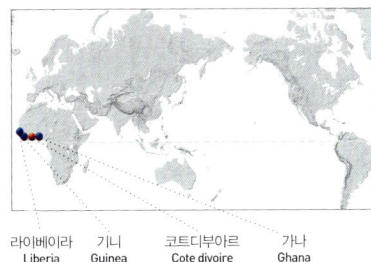

라이베리아 / Liberia  기니 / Guinea  코트디부아르 / Cote divoire  가나 / Ghana

♀
코트디부아르 산 70mm
(Akupe, Cote divoire. 2005. 5.)

♂
코트디부아르 산 67mm
(Akupe, Cote divoire. 2006. 4.)

# 꽃무지아과 / Cetoniinae

## Fornasinius

| 학 명 Scientific name | 채집국 Collected locality | 크 기 Size | 참 고 Remark |
|---|---|---|---|
| 포르나시니왕꽃무지<br>*Fornasinius fornasinii* Bertoloni, 1852 |  탄자니아<br>Tanzania | ♂ 40-70mm, ♀ 35-60mm | Subtribe Goliathina |

중앙아프리카 남동부에 서식한다. 몸빛은 부드러운 광택을 지니며 검정색 바탕에 황색의 띠가 둘러져 있다. 가슴은 정중앙으로 한 줄의 황색 세로줄을 가지고 있으며 딱지날개는 세로줄의 황색 점무늬들이 있다. 하나의 머리뿔말단부는 두 갈래로 뾰족하게 갈고리 모양으로 발달하였고 머리기저부 양쪽에 가시처럼 발달한 한 쌍의 돌기가 있다.

Inhabiting southeastern part of Central Africa, they have a softly-glossy black body which is yellow-striped. On the center of the thorax runs a pair of yellow stripes while the elytra (hard wings) show some stripes of yellow speckles. Its cephalic horn is hooked and pointed, two-forked at the tip. A pair of protrusions like thorns can be seen at both sides of the lower cephalic part.

[Distribution] Eastern Uganda, Western Kenya, Eastern D.R. Congo, Burundi, Rwanda, Tanzania, Mozambique

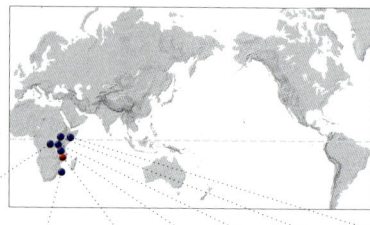

♂ 탄자니아 산 48mm
(Sanya, Tanzania. 2003. 3.)

♀ 탄자니아 산 47mm
(Sanya, Tanzania. 2003. 3.)

Flower Beetles / Tribe 5. Goliathini

## Fornasinius

| 학 명 Scientific name | 채집국 Collected locality | 크 기 Size | 참 고 Remark |
|---|---|---|---|
| 루쑤스왕꽃무지<br>*Fornasinius russus* Kolbe, 1884 | 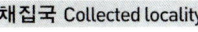 우간다<br>Uganda | ♂ 40-70mm, ♀ 35-60mm | Subtribe Goliathina |

중앙아프리카 중부에 서식한다. 몸빛은 부드러운 광택을 지니며 적갈색 바탕에 검정색의 띠가 둘러져 있다. 하나의 머리뿔말단부는 두 갈래로 뾰족하게 갈고리 모양으로 발달하였고 *F. fornasinii*와 비교하여 머리기저부에 양쪽으로 가시모양의 돌기가 방패형으로 더욱 크게 발달하였다.

They inhabit the middle region of Central Africa. Their dark-brown body is softly glossy and surrounded by a black stripe. Its cephalic horn is hooked and pointed, two-forked at the tip. The pointed and shield-shaped protrusions on both sides of the lower cephalic part are bigger than those of *F. fornasinii*.

[Distribution] Congo, Gabon, Northern D.R. Congo, Western Uganda

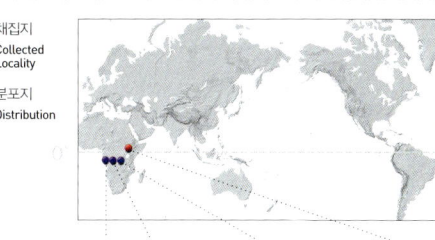

♀

우간다 산 54mm
(Uganda. 2004. 9.)

♂

우간다 산 59mm
(Uganda. 2004. 9.)

## 꽃무지아과 / Cetoniinae

### Hypselogenia

| 학 명 Scientific name | 채집국 Collected locality | 크 기 Size | 참 고 Remark |
|---|---|---|---|
| 코로사얼룩뿔꽃무지<br>*Hypselogenia corrosa* Bates, 1881 | 탄자니아<br>Tanzania | ♂ 18-28mm, ♀ 23-27mm | Subtribe Ichnestomina |

중앙아프리카 중동부에 서식한다. 몸빛은 검정바탕에 가슴은 황백색의 띠가 양쪽으로 둘러져 있으며 위, 아래로 한 쌍씩의 반점을 가지고 있다. 딱지날개는 황백색의 띠와 반점들을 가지고 있다. 하나의 머리뿔말단부는 두 갈래로 나뉘어져 있으나 크게 발달하지는 않았다.

Inhabiting the middle region of Central Africa, they have a black body surrounded by yellowish-white stripes with a pair of speckles at the top and bottom. The elytra (hard wings) show some yellowish-white stripes and speckles. Its cephalic horn is two-forked at the tip, though not so developed.

[Distribution] Northern Tanzania, Kenya, Uganda

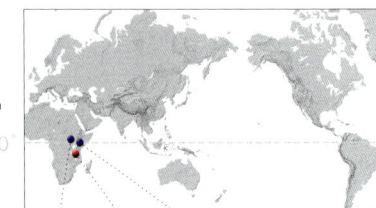

♂

탄자니아 산 25mm
(Sanya, Tanzania. 2003. 3.)

♀

탄자니아 산 23mm
(Sanya, Tanzania. 2003. 3.)

# Flower Beetles / Tribe 5. Goliathini

## Dicranocephalus

| 학 명 Scientific name | 채집국 Collected locality | 크 기 Size | 참 고 Remark |
|---|---|---|---|
| 왈라치사슴풍뎅이<br>*Dicranocephalus wallichii* Hope, 1832 | 타이 Thailand | ♂ 24-40mm, ♀ 22-26mm | Subtribe Dicronocephalina |

서남아시아, 인도차이나 반도에 서식한다. 몸빛은 광택을 지닌 적갈색바탕에 머리와 가슴 딱지날개는 마치 벨벳과 같이 황색의 미모로 덮여있다. 가슴에는 한 쌍의 세로줄을 가지고 있으나 끝까지 연결되어 있지는 않다. 딱지날개 위쪽의 양쪽과 아래에 각각 한 쌍의 적갈색 반점들을 가지고 있다. 머리뿔은 기저부에서부터 두 갈래로 나뉘어져 마치 수사슴의 뿔과 닮아있다.

They inhabit the southwestern parts of Asia and the Indochinese Peninsula. Their body is glossy and reddish-brown while the cephalic part, thorax and elytra (hard wings) are covered with velvet-like fine yellow hair. There is a pair of stripes on the thorax, which gets blurry at the tips. On the upper side and lower side of the elytra (hard wings) is a pair of reddish-brown speckles. The cephalic horn, divided into two from the lower part, resembles an antler.

[Distribution] India, Nepal, Bhutan, Tibet, Myanmar, Thailand, Laos, Vietnam

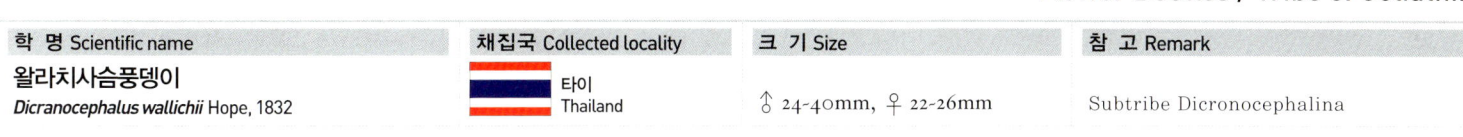

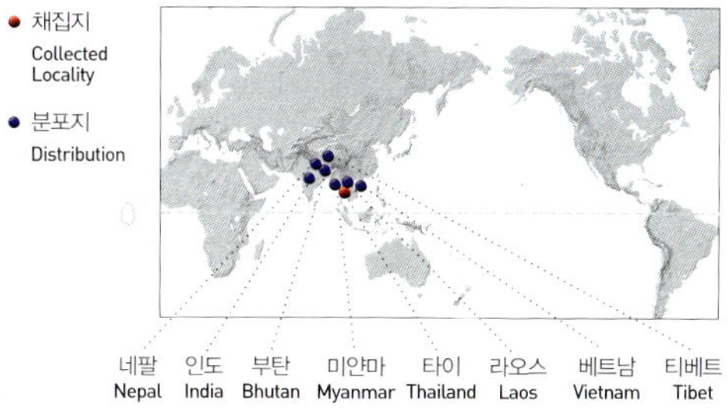

♂

타이 산36mm
(Chiang-Mai, N.Thailand. 2002. 3.)

## 꽃무지아과 / Cetoniinae

**Dicarnocephalus**

| 학 명 Scientific name | 채집국 Collected locality | 크 기 Size | 참 고 Remark |
|---|---|---|---|
| 사슴풍뎅이<br>*Dicarnocephalus adamsi* Pascoe, 1863 | 대한민국 Korea | ♂ 23-38mm, ♀ 20-25mm | Subtribe Dicronocephalina |

한반도, 중국, 티벳 동부에 서식한다. 몸빛은 약한 광택을 지닌 적갈색바탕에 머리와 가슴 딱지날개는 마치 벨벳과 같이 황색의 미모로 덮여있다. 가슴에는 한 쌍의 세로줄을 가지고 있으나 끝까지 연결 되지는 않고, *D. wallichii*와 비교하여 세로줄의 띠가 더 넓고 양측에 한 쌍의 작은 반점을 가지고 있다. 딱지날개 위쪽의 양쪽과 아래에 각각 한 쌍의 적갈색 반점들을 가지고 있다. 머리뿔은 기저부에서부터 두 갈래로 나뉘어져 마치 수사슴의 뿔과 닮아있다. 서식지와 개체의 크기에 따라 머리뿔의 모양에 변화가 많다. 성충은 5월~6월초에 가장 많이 출현하며 그룹을 이루어 짝짓기 하는 모습을 관찰할 수 있다.

♂

전북 정읍 산 27mm
(Mt. Naejangsan, Jeong-eup, Jeollabuk-do, S. Korea. 1988. 5.)

♀

전남 담양 산 21mm
(Mt. Baekyangsan, Danyang-gun, Jeollabuk-do, S. Korea. 1988. 5.)

# Flower Beetles / Tribe 5. Goliathini

They inhabit the Korean Peninsula, China and the eastern part of Tibet. Their body is slightly glossy and reddish-brown while the cephalic part, thorax and elytra (hard wings) are covered with velvet-like fine yellow hair. There is a pair of stripes on the thorax, which gets blurry at the tips. The stripes, having a pair of small speckles on both sides, are wider than those of *D. wallichii*. On each of both upper sides and lower side of the elytra (hard wings) is a pair of reddish-brown speckles. The cephalic horn, divided into two from the lower part, resembles an antler. The shape of the cephalic horns depends heavily on where the habitat is and how big a specimen is. The biggest number of adult insects appears in May to early June, mating in groups.

[Distribution] Korea, China, Tibet

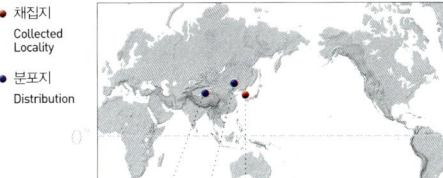

전북 부안 산 30mm
(Mt. Byeonsan, Buan-gun, Jeollabuk-do, S. Korea. 1987. 5.)

## 꽃무지아과 / Cetoniinae

**Cheirolasia**

| 학 명 Scientific name | 채집국 Collected locality | 크 기 Size | 참 고 Remark |
|---|---|---|---|
| 부르케이앞장다리꽃무지<br>*Cheirolasia burkei burkei* Westwood, 1843 | 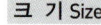 짐바브웨<br>Zimbabwe | ♂ 26-34mm, ♀ 23-27mm | Subtribe Coryphocerina |

남아프리카 중동부에 서식한다. 이 종은 *C. burkei*를 원아종을 포함하여 *C. b. Septentrionis, C. b. lettowvorbecki, C. b. hopei, C. b. histrio, C. b. burkei* 여섯 아종이 있다. 이 종의 몸빛은 암적색바탕에 가슴은 백색의 두터운 테두리와 딱지날개에 백색의 반점을 가지고 있다. 암컷은 황색이다. 머리뿔은 기저부에서 세 개로 나뉘어져 있으며 한 쌍은 가시처럼 발달하였고 가운데 뿔은 반달형 돌칼모양을 하고 있다.

This species, which inhabits the mid-eastern region of South Africa, has six subspecies: *C. burkei, C. b. Septentrionis, C. b. lettowvorbecki, C. b. hopei, C. b. histrio, C. b. burkei*. Its body is dark-red and the thorax has a thick and white outline while the elytra (hard wings) have some white speckles. Its females have a yellow body. The cephalic horn is divided into three from the lower part, with the middle one resembling a crescent knife and the rest developed like thorns.

[Distribution] Rep. South Africa, Zimbabwe

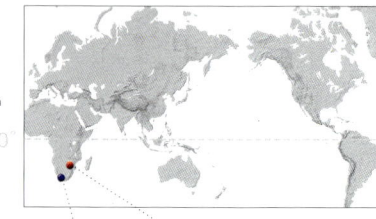

짐바브웨 산 31mm
(Bulawayo, Zimbabwe. 2003. 7.)

짐바브웨 산 27mm
(Bulawayo, Zimbabwe. 2003. 7.)

# Flower Beetles / Tribe 5. Goliathini

## Cheirolasia

| 학 명 Scientific name | 채집국 Collected locality | 크 기 Size | 참 고 Remark |
|---|---|---|---|
| 부르케이셉텐트리오니스앞장다리꽃무지<br>*Cheirolasia burkei septentrionis* Kriesche, 1921 |  탄자니아<br>Tanzania | ♂ 29-32mm, ♀ 24-28mm | Subtribe Coryphocerina |

중앙아프리카에 서식한다. 이 아종은 *C. b. burkei*에 비하여 밝은 노랑색을 띠하고 있다. *C. b. burkei*와 함께 수컷의 앞다리는 길게 발달하여 종아리마디와 발목마디에 솔과 같은 황색의 강모를 가지고 있다. 머리뿔은 *C. b. burkei*와 비교하여 두텁다.

This subspecies, which inhabits Central Africa, is more yellow and brighter than *C. b. burkei*. The males' forelegs, long and well-developed, have some yellow and brush-like bristles on the thigh and ankle joints like those of *C. b. burkei*. The cephalic horn is thicker than that of *C. b. burkei*.

[Distribution] D.R. Congo, Kenya, Tanzania

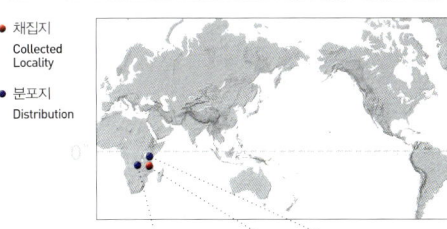

♀

탄자니아 산 28mm
(Mt. Uluguru, Tanzania. 2003. 7.)

♂

탄자니아 산 31mm
(Mt. Uluguru, Tanzania. 2003. 7.)

## 꽃무지아과 / Cetoniinae

**Dicronorrhina**

| 학 명 Scientific name | 채집국 Collected locality | 크 기 Size | 참 고 Remark |
|---|---|---|---|
| 데르비아나왕꽃무지<br>*Dicronorrhina derbyana derbyana* Westwood, 1843 | 짐바브웨<br>Zimbabwe | ♂ 31-48mm, ♀ 28-44mm | Subtribe Coryphocerina |

아프리카 중남부에 넓게 서식한다. 이 종은 다섯 아종이 있으며 머리가 방패모양으로 발달하여 두 갈래로 갈라진 짧은 뿔을 가지고 있다. 몸빛은 광택을 가진 녹색 바탕에 백색의 줄무늬를 가지고 있다. 암컷은 광택이 보다 강하다.

This species, has five subspecies. Which inhabits the mid-to-southern parts of Central Africa, has a shield-shaped cephalic part with a short, two-forked horn. Its body is green and glossy with some white stripes. The females are glossier than the males.

[Distribution] Southern Tanzania, Mozambique, Malawi, Southeastern D.R. Congo, Zambia, Rec. of South America, Angola, Botswana,

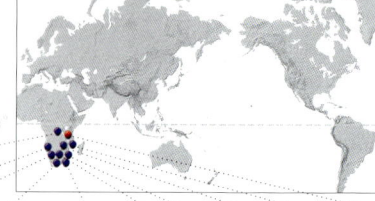

♂

짐바브웨 산 48mm
(Guruve, Mashonaland, Zimbabwe. 2003. 5.)

♀

짐바브웨 산 44mm
(Guruve, Mashonaland, Zimbabwe. 2003. 5.)

Flower Beetles / Tribe 5. Goliathini

## Dicronorrhina

| 학 명 Scientific name | 채집국 Collected locality | 크 기 Size | 참 고 Remark |
|---|---|---|---|
| 데르비아나콘라드씨왕꽃무지<br>*Dicronorrhina derbyana conradsi* | 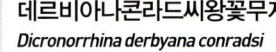 탄자니아<br>Tanzania | ♂ 31-47mm, ♀ 28-44mm | Subtribe Coryphocerina |

중앙아프리카 동부에 서식한다. 이 아종은 머리가 방패모양으로 발달하여 두 갈래로 갈라진 짧은 뿔을 가지고 있다. 몸빛은 광택을 가진 적갈색 바탕에 백색의 줄무늬를 가지고 있다. 암컷은 짙은 적갈색이며 광택이 보다 강하다.

This subspecies, which inhabits the eastern part of Central Africa, has a shield-shaped cephalic part with a short, two-forked horn. Its body is dark-red and glossy with some white stripes. The females have a dark reddish-brown body, glossier than the males.

[Distribution] Tanzania

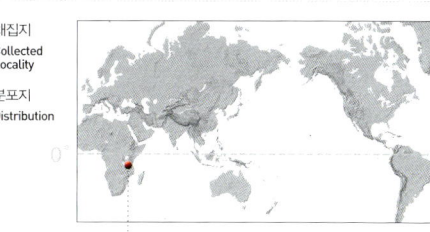

● 채집지 Collected Locality
● 분포지 Distribution

탄자니아
Tanzania

♀

탄자니아 산 33mm
(Usambarae, Tanzania. 2004. 11.)

♂

탄자니아 산 38mm
(Usambarae, Tanzania. 2004. 11.)

## 꽃무지아과 / Cetoniinae

### Dicronorrhina

| 학 명 Scientific name | 채집국 Collected locality | 크 기 Size | 참 고 Remark |
|---|---|---|---|
| 데르비아나오벨투에리왕꽃무지<br>*Dicronorrhina derbyana oberthueri* Deyroll, 1876 | 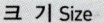 탄자니아 Tanzania | ♂ 29.5-49mm, ♀ 32-44.5mm | Subtribe Coryphocerina |

중앙아프리카 동부에 서식한다. 이 종은 머리가 방패모양으로 발달하여 두 갈래로 갈라진 짧은 뿔을 가지고 있다. 몸빛은 광택을 가진 황록색을 띠며 가슴의 중앙부는 황적색을 띤다. 암컷은 광택이 보다 강하다.

This species, which inhabits the eastern part of Central Africa, has a shield-shaped cephalic part with a short, two-forked horn. Its body is yellowish-green and glossy while the center of the thorax is yellowish-red. The females are glossier than the males.

[Distribution] Somalia, Uganda, Kenya, Tanzania

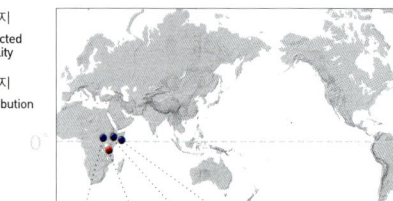

♂

탄자니아 산 44mm
(E-Mt. Usambarae, Tanzania. 2004. 6.)

♀

탄자니아 산 39mm
(E-Mt. Usambarae, Tanzania. 2004. 6.)

# Flower Beetles / Tribe 5. Goliathini

## Eudicella

| 학 명 Scientific name | 채집국 Collected locality | 크 기 Size | 참 고 Remark |
|---|---|---|---|
| 큐프레오수투랄리스뿔꽃무지<br>*Eudicella cupreosuturalis* Bourgoin, 1913 | 카메룬<br>Cameroon | ♂ 32-40mm, ♀ 30-34mm | Subtribe Coryphocerina |

중앙아프리카에 서식한다. *Eudicella*속은 20여종이 알려져 있다. 이 속은 다채로운 색상과 줄무늬를 가진 것이 많다. 이 종의 머리뿔은 두 갈래로 갈라져 짧게 위로 굽어있으며 기저부의 가시와 같은 한 쌍의 돌기는 전방을 향하여 있다. 광택을 지니며 앞다리넓적마디에 황색의 강모를 가지고 있다.

The *Eudicella* genus, inhabiting Central Africa, is divided into 20 or so species, most of which have various colors and stripes. The two-forked cephalic horn of this species is bent upward and short, while a pair of protrusions in the lower part, resembling thorns, points forward. Its glossy body and or the frontal thigh(femur) have some yellow bristles.

[Distribution] Cameroon, Northern D.R. Congo, Burundi, Uganda, Kenya

● 채집지 Collected Locality
● 분포지 Distribution

카메룬 Cameroon | 콩고민주공화국(자이르) 북부 N. D. R. Congo(Zaire) | 부룬디 Burundi | 우간다 Uganda | 케냐 Kenya

♂

카메룬 산 33mm
(Mt. Cmeroon, Cmeroon. 2005. 6.)

## 꽃무지아과 / Cetoniinae

**Eudicella**

| 학 명 Scientific name | 채집국 Collected locality | 크 기 Size | 참 고 Remark |
|---|---|---|---|
| 그랄리뿔꽃무지<br>*Eudicella gralli gralli* (Buquet, 1836) |  콩고<br>Congo | ♂ 32-44mm, ♀ 28-33mm | Subtribe Coryphocerina |

중앙아프리카에 넓게 서식한다. 이 종의 머리뿔은 두 갈래로 갈라져 마치 물소뿔처럼 발달하였다. 기저부의 가시와 같은 한 쌍의 돌기는 전방을 향하고 있다. 광택을 지니며 앞다리종아리마디 안쪽에 가시와 같은 돌기를 가지고 있다.

The two-forked cephalic horn of this species, which has a wide habitat across Central Africa, resembles that of a water buffalo. A pair of thorn-like protrusions on the lower part pointss forward. It has a glossy body and some thorn-like protrusion inside the frontal calf joints.

[Distribution] Central Africa Rep., Gabon, Congo, D.R. Congo

● 채집지 Collected Locality
● 분포지 Distribution

가봉 Gabon / 콩고 Congo / 콩고민주공화국(자이르) D.R. Congo(Zaire) / 중앙아프리카공화국 Central Africa Rep.

♂

콩고 산 35mm
(PK. Rouge, Congo. 2003. 4)

♀

콩고 산 30mm
(PK. Rouge, Congo. 2003. 4)

# Flower Beetles / Tribe 5. Goliathini

## Eudicella

| 학 명 Scientific name | 채집국 Collected locality | 크 기 Size | 참 고 Remark |
|---|---|---|---|
| 그랄리움부로비타타뿔꽃무지<br>*Eudicella gralli umbrovittata* Allard, 1985 | 케냐<br>Kenya | ♂ 31-53mm, ♀ 27-34mm | Subtribe Coryphocerina |

중앙아프리카 동부에 서식한다. 이 종의 머리뿔은 *E. g. gralli*와 비교하여 보다 크고 길게 발달하여 기저부의 가시와 같은 한 쌍의 돌기는 45° 각도로 벌어져 있다. 광택을 지니며 앞다리종아리마디 안쪽에 톱날과 같은 돌기를 가지고 있다.

This species, which inhabits the eastern part of Central Africa, has a cephalic horn bigger and longer than that of *E. g. gralli*. A pair of thorn-like protrusions on the lower part forms an oblique angle. It has a glossy body and some thorn-like protrusion inside the frontal calf joints.

[Distribution] Uganda , Kenya

우간다  케냐
Uganda Kenya

♀

케냐 산 33mm
(W. Kenya. 2003. 5)

♂

케냐 산 46mm
(W. Kenya. 2003. 5)

# 꽃무지아과 / Cetoniinae

## Eudicella

| 학 명 Scientific name | 채집국 Collected locality | 크 기 Size | 참 고 Remark |
|---|---|---|---|
| 스미씨쉬라티카뿔꽃무지<br>*Eudicella smithi shiratica* Csiki, 1909 |  콩고<br>Congo | ♂ 30-41mm, ♀ 28-32mm | Subtribe Coryphocerina |

중앙아프리카 중동부에 서식한다. 이 종의 머리뿔은 두 갈래로 갈라져 안쪽으로 굽어있다. 기저부의 가시와 같은 한 쌍의 돌기는 전방을 향하고 있다. 광택을 지니며 앞다리종아리마디 안쪽에 가시와 같은 돌기를 가지고 있다. *Eudicella*속 수컷의 뿔은 몸 크기와 개체에 따라 변이가 있다.

The cephalic horn of this species, which inhabits the mid-to-eastern parts of Central Africa, is divided into two and bent inward. A pair of thorn-like protrusions on the lower part pointss forward. It has a glossy body and some thorn-like protrusion inside the frontal calf joints. Horns of the *Eudicella* males show some variations depending on body size and individual specimens.

[Distribution] D.R. Congo, Uganda, Kenya

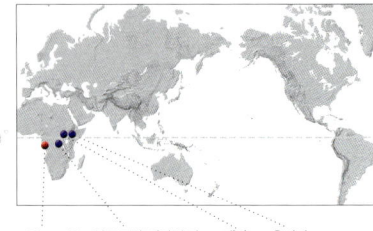

콩고 / Congo   콩고민주공화국(자이르) / D.R. Congo(Zaire)   케냐 / Kenya   우간다 / Uganda

♂
콩고 산 40mm
(Congo. 2003. 4)

♀
콩고 산 31mm
(Congo. 2003. 4)

# Flower Beetles / Tribe 5. Goliathini

## Eudicella

| 학 명 Scientific name | 채집국 Collected locality | 크 기 Size | 참 고 Remark |
|---|---|---|---|
| 트릴리네아뿔꽃무지<br>*Eudicella trilineata* Quedenfeldt, 1880 | 말라위<br>Malawi | ♂ 26-37mm, ♀ 25-33mm | Subtribe Coryphocerina |

중앙아프리카중동부에 서식한다. 이 종의 *E. s. shiratica*와 비교하여 머리뿔은 작고 두 갈래로 갈라져 안쪽으로 굽어있다. 기저부의 돌기는 방패모양으로 발달하였다. 광택을 지니며 앞다리종아리마디 안쪽에 가시와 같은 돌기를 가지고 있다. 딱지날개 위아래 양쪽에 검은 반점을 가지고 있다.

The two-forked and inward-bent cephalic horn of this species, which inhabits the mid-to-eastern parts of Central Africa, is smaller than that of *E. s. shiratica*. The protrusion on the lower part resembles a shield. It has a glossy body and some thorn-like protrusion inside the frontal calf joints(tibia). Some black speckles are visible underneath and on both sides of the elytra (hard wings).

[Distribution] Southern Tanzania, Malawi

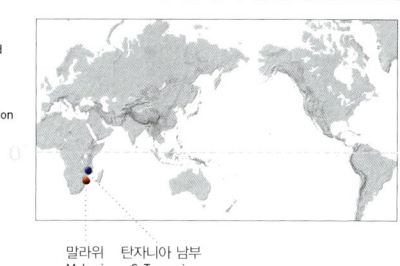

♀

말라위 산 33mm
(Malawi. 2003. 5.)

♂

말라위 산 39mm
(Malawi. 2003. 5.)

## 꽃무지아과 / Cetoniinae

### Ingrisma

| 학 명 Scientific name | 채집국 Collected locality | 크 기 Size | 참 고 Remark |
|---|---|---|---|
| 유리르히나붉은다리주걱턱꽃무지<br>*Ingrisma euryrrhina* (Gestro, 1891) | 타이 Thailand | ♂♀ 23-30mm | Subtribe Coryphocerina |

인도차이나 반도 북서부에 서식한다. *Ingrisma*속은 12여종이 알려져 있다. 이종의 몸빛은 광택을 가진 검정색이며 종아리마디들이 적갈색을 띤다. 머리 돌기는 방패형으로 마치 하마 머리처럼 발달하였다.

This species inhabits the northwestern region of the Indochinese Peninsula. Currently, approximately 12 species fall under the *Ingrisma* genus. This species has a black glossy body and each calf joint(tibia) is reddish-brown. The shield-shaped cephalic protrusion is developed like the head of a hippopotamus.

[Distribution] Myanimar. Thailand

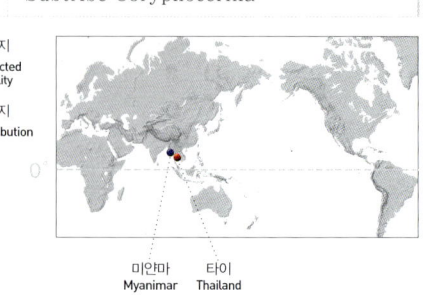

♀

타이 북부 산 27mm
(Chiang-Mai, N. Thailand. 2004. 11.)

♀

타이 북부 산 26mm
(Chiang-Mai, N. Thailand. 2004. 11.)

# Flower Beetles / Tribe 5. Goliathini

## Cyphonocephalus

| 학 명 Scientific name | 채집국 Collected locality | 크 기 Size | 참 고 Remark |
|---|---|---|---|
| 오리바세우스사슴뿔꽃무지<br>*Cyphonocephalus olivaceus* (Dupont, 1835) | 인도 India | ♂ 30-34mm, ♀ 22-26mm | Subtribe Coryphocerina |

인디아 남부에 서식한다. 이 종은 1속 1종이다. 몸빛은 광택을 지닌 밝은 녹색으로 머리뿔과 종아리마디 아래로는 적갈색이다. 머리의 뿔은 마치 젊은 사슴의 뿔을 닮았다.

This species, which inhabits the northern and southern regions of India, constitutes a genus alone. It has a bright-green, glossy body and the cephalic horn is reddish-brown like the lower part of the calf joints(tibia). The cephalic horn resembles the antler of a young deer

[Distribution] Southern India

● 채집지 Collected Locality
● 분포지 Distribution

인도 남부
S. India

♂

인도 남부 산 30mm
(Nilgiri Hills, India. 2004. 11.)

## 꽃무지아과 / Cetoniinae

| 학 명 Scientific name | 채집국 Collected locality | 크 기 Size | 참 고 Remark |
|---|---|---|---|
| 훼레로이미네티큐앞장다리꽃무지<br>*Jumnos ferreroiminettiique* Antoine, 1991 | 타이 Thailand | ♂ 42-51mm, ♀ 40-45mm | Subtribe Coryphocerina |

인도차이나 반도 북부에 서식한다. *Jumnos*속은 중대형에 속하는 꽃무지이다. 이종의 몸빛은 광택이 강한 암녹색의 금속형 광택을 띤다. 빛의 방향에 따라 그 색깔도 따라 바뀐다. 앞다리가 길게 발달하여 있다.

This species inhabits the northern region of the Indochinese Peninsula. The *Jumnos* genus is a relatively big-sized group of Cetoniid beetles. This species has a dark-green body with strong metallic gloss, which displays different colors depending on the direction of light. Its forelegs are long and well-developed.

[Distribution] Thailand, Laos

♂

타이 북부 산 51mm
(Chiang-Mai, N. Thailand. 2004. 6.)

♀

타이 북부 산 41mm
(Chiang-Mai, N. Thailand. 2004. 6.)

Flower Beetles / Tribe 5. Goliathini

## Jumnos

| 학 명 Scientific name | 채집국 Collected locality | 크 기 Size | 참 고 Remark |
|---|---|---|---|
| 룩케리노랑네점박이앞장다리꽃무지<br>*Jumnos ruckeri* Saunders, 1839 | 타이<br>Thailand | ♂ 38-56mm, ♀ 38-46mm | Subtribe Coryphocerina |

인도차이나 반도 북부에 서식한다. 이 종은 *J. r. ruckeri*, *J. r. tonkinensis*, *J. r. pfanneri*와 함께 3아종으로 나뉜다. 이 아종은 녹색의 금속성 광택을 띤다. 딱지날개는 황색의 네 개의 큰 반점을 가지고 있다. *J. r. pfanneri* 만이 황색의 반점이 없거나 매우 작다.

This species, which has some metallic green gloss, inhabits the northern region of the Indochinese Peninsula. This subspecies, with *J. r. ruckeri*, *J. r. tonkinensis* and *J. r. pfanneri*, is grouped into three subspecies. *J. r. pfanneri* has very small or no yellow speckles on its elytra (hard wings) while four yellow and big speckles are seen on those of the others.

[Distribution] Northastern India, Myanimar, Thailand, Malay Peninsula, Vietnam

♀
타이 북부 산 46mm
(Chiang-Mai, N. Thailand. 2003. 7.)

♂
타이 북부 산 52mm
(Chiang-Mai, N. Thailand. 2003. 7.)

# 꽃무지아과 / Cetoniinae

## Mecynorhina

| 학 명 Scientific name | 채집국 Collected locality | 크 기 Size | 참 고 Remark |
|---|---|---|---|
| 크라치지줄무늬귀신꽃무지<br>*Mecynorhina kraatzi* Moser, 1905 |  카메룬<br>Cameroon | ♂ 38-70mm, ♀ 40-49mm | Subtribe Coryphocerina |

아프리카 중서부에 서식한다. *Mecynorhina*속의 국명은 2002년도에 발간된 "세계 곤충도감"에서 저자 홍승표씨로 부터 귀신꽃무지로 명명 되어졌다. 이 종의 머리뿔은 크고 길게 발달하였으며 말단부는 두 갈래로 갈라져 있다. 머리 기저부의 가시와 같은 돌기는 작고 45° 각도로 벌어져 있다. 앞다리종아리마디에 돌기가 큰 톱날처럼 발달하였다.

This species inhabits the midwestern part of Africa. The local name of the *Mecynorhina* genus *Mecynorhina torquata* was first used in 2002 by Seung-Pyo Hong, when Insects of the World was published. This species has a long and well-developed cephalic horn, which is two-forked at the tip. A pair of thorn-like protrusions on the lower part forms an oblique angle. Some protrusions are well-developed like saw blades on the frontal calf joints(protibia).

[Distribution] Nigeria, Cameroon

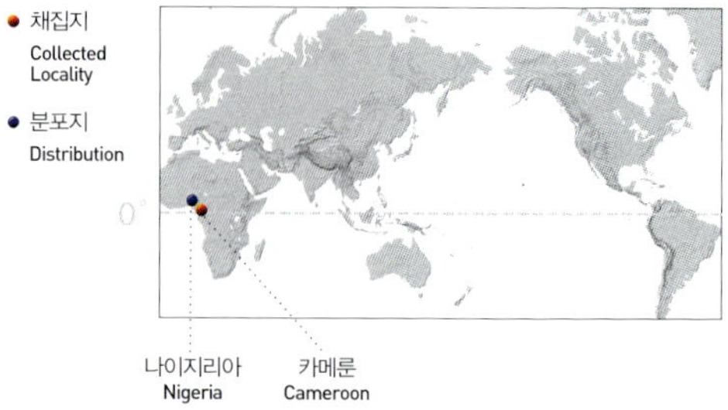

♂  
카메룬 산 60mm  
(Mt. Cameroun, Cameroon. 2005. 2.)

♀  
카메룬 산 49mm  
(Mt. Cameroun, Cameroon. 2005. 2.)

Flower Beetles / Tribe 5. Goliathini

## Mecynorhina

| 학 명 Scientific name | 채집국 Collected locality | 크 기 Size | 참 고 Remark |
|---|---|---|---|
| 사바게이점박이귀신꽃무지<br>*Mecynorhina savagei* Moser, 1905 | 카메룬<br>Cameroon | ♂ 38-70mm, ♀ 40-52mm | Subtribe Coryphocerina |

아프리카 중서부에 서식한다. 이 종의 머리뿔은 크고 길게 발달하였으며 말단부는 두 갈래로 갈라져 있다. 머리 기저부의 돌기는 작은 뿔로 발달하여 45° 각도로 벌어져 안쪽으로 굽어져 있다. 앞다리종아리마디에 돌기가 큰 톱날처럼 발달하였다.

This species inhabits the midwestern part of Africa. The cephalic horn of this species is big and long with its tip divided into two. The protrusions on the lower cephalic part, like small horns, form an oblique angle and are bent inward. Some protrusions are well-developed like saw blades on the frontal calf joints(protibia).

[Distribution] Nigeria, Cameroon

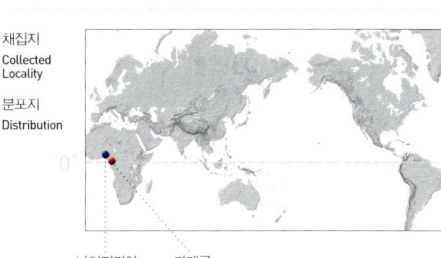

♀

카메룬 산 52mm
(Kumba, Cameroon. 2005. 2.)

♂

카메룬 산 60mm
(Kumba, Cameroon. 2005. 2.)

# 꽃무지아과 / Cetoniinae

## Mecynorhina

| 학 명 Scientific name | 채집국 Collected locality | 크 기 Size | 참 고 Remark |
|---|---|---|---|
| 주홍대왕귀신꽃무지 (유니칼라 형)<br>*Mecynorhina oberthueri* Fairmaire, 1903 -Form unicolor |  탄자니아<br>Tanzania | ♂ 47-74mm, ♀ 45-64mm | Subtribe Coryphocerina |

중앙아프리카 동부에 서식한다. 이 종의 머리뿔은 머리방패와 함께 발달하였고 말단부는 갈라져 있지 않다. 머리 기저부의 방패와 같은 돌기는 전방향으로 발달하였다. 앞다리종아리마디에 난 돌기는 큰 톱날처럼 발달하였다.

The cephalic horn of this species, which inhabits the eastern region of Central Africa, is developed with its cephalic shield and not forked in the tip. The shield-like protrusion on the lower cephalic part is pointed forward. Some protrusions are well-developed like saw blades on the frontal calf joints(protibia).

[Distribution] Northestern Tanzania

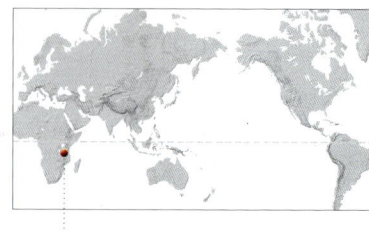

● 채집지 Collected Locality
● 분포지 Distribution

탄자니아 북동부
NE. Tanzania

♂

탄자니아 산 68mm
(Mt. Zengia, Tanzania. 2005. 3.)

♀

탄자니아 산 54mm
(Mt. Zengia, Tanzania. 2005. 3.)

Flower Beetles / Tribe 5. Goliathini

## Mecynorhina

| 학 명 Scientific name | 채집국 Collected locality | 크 기 Size | 참 고 Remark |
|---|---|---|---|
| 주홍점박이대왕귀신꽃무지 (데코라타 형)<br>*Mecynorhina oberthueri* Fairmaire, 1903 - **Form decorata** |  탄자니아<br>Tanzania | ♂ 47-74mm, ♀ 45-64mm | Subtribe Coryphocerina |

중앙아프리카 동부에 서식한다. 이 종의 머리뿔은 머리방패와 함께 발달하였고 말단부는 갈라져 있지 않다. 머리 기저부의 방패와 같은 돌기는 전방향으로 발달하였다. 앞다리종아리마디에 돌기가 큰 톱날처럼 발달하였다. *M. o. unicolor*와 비교하여 적황색반점의 무늬를 가지고 있다.

The cephalic horn of this species, which inhabits the eastern region of Central Africa, is developed with its cephalic shield and not forked in the tip. The shield-like protrusion on the lower cephalic part is pointed forward. Some protrusions are well-developed like saw blades on the frontal calf joints(protibia). It is different from *M. o. unicolor* in that it has some patterns of reddish-yellow speckles.

[Distribution] Northestern Tanzania

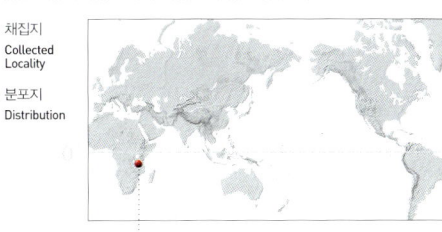

탄자니아 북동부
NE. Tanzania

♀

탄자니아 산 53mm
(N. Kive Lake, Tanzania. 2006. 8.)

♂

탄자니아 산 60mm
(N. Kive Lake, Tanzania. 2006. 8.)

## 꽃무지아과 / Cetoniinae

**Mecynorhina**

| 학 명 Scientific name | 채집국 Collected locality | 크 기 Size | 참 고 Remark |
|---|---|---|---|
| 토르콰타-이마쿠리콜리스대왕귀신꽃무지<br>*Mecynorhina torquata immaculicollis* (kraatz, 1890) |  카메룬<br>Cameroon | ♂ 58-77mm, ♀ 47-58mm | Subtribe Coryphocerina |

중앙아프리카 동부에 서식한다. 이 종은 *M. t. torquata, M. t. immaculicollis, M. t. poggei, M. t. ugandensis* 4아종이 알려져 있으며, 머리방패는 마름모형이며 뿔은 하나로 길고 크게 발달하여 위쪽으로 굽어있으며 말단부에 돌기를 가지고 있다. 앞다리종아리마디의 돌기가 큰 톱날처럼 발달하였다.

This species inhabits the eastern part of Central Africa. *M. torquata* is divided into four subspecies: *M. t. torquata, ugandensis, M. t. immaculicollis , M. t. poggei* and *M. t. ugandensis*. The cephalic shield of the species is lozenge-shaped and the single, long and big horn is bent upward with some protrusions at the tip. Some protrusions are well-developed like saw blades on the frontal calf joints(protibia).

[Distribution] Northeastern Tanzania, Southern Central African. Rep., Gabon, Congo, D.R. Congo

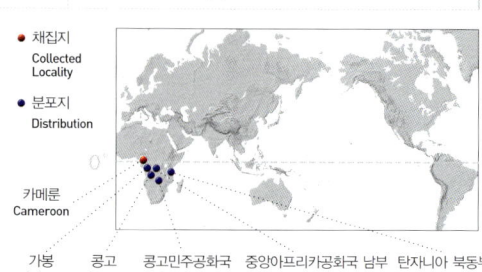

♂

카메룬 산 77mm
(Kupe, Cameroon. 2006. 8.)

♀

카메룬 산 52mm
(Kupe, Cameroon. 2006. 8.)

Flower Beetles / Tribe 5. Goliathini

## Mecynorhina

| 학 명 Scientific name | 채집국 Collected locality | 크 기 Size | 참 고 Remark |
|---|---|---|---|
| 토르콰타-포게이대왕귀신꽃무지<br>*Mecynorhina torquata poggei* (kraatz, 1890) | 콩고민주공화국<br>D. R. Congo | ♂ 55-86mm, ♀ 45-59mm | Subtribe Coryphocerina |

중앙아프리카 동부에 서식한다. 이종의 머리방패는 마름모형이며 뿔은 하나로 길고 크게 발달하여 위쪽으로 굽어있으며 말단부에 돌기는 *M. t. immaculicollis*와 비교하여 보다 크며 가슴과 딱지날개에 황백색의 줄무늬를 가지고 있다. 앞다리종아리마디의 돌기가 큰 톱날처럼 발달하였다.

This species inhabits the eastern part of Central Africa. The cephalic shield of the species is lozenge-shaped and the single, long and big horn is bent upward while the protrusion at the tip is bigger than that of *M. t. immaculicollis*. It has some yellowish-white stripes on the thorax and elytra (hard wings). Some protrusions are well-developed like saw blades on the frontal calf joints(protibia).

[Distribution] Southeastern D.R. Congo

콩고민주공화국(자이르)
D. R. Congo(Zaire)

♀

콩고민주공화국 산 58mm
(Khoni(Shaba), D. R. Congo. 2006. 1.)

♂

콩고민주공화국 산 86mm
(Khoni(Shaba), D. R. Congo. 2006. 1.)

## 꽃무지아과 / Cetoniinae

**Mecynorhina**

| 학 명 Scientific name | 채집국 Collected locality | 크 기 Size | 참 고 Remark |
|---|---|---|---|
| 토르콰타-우간덴시쓰대왕귀신꽃무지<br>*Mecynorhina torquata ugandensis* (kraatz, 1907) |  콩고민주공화국<br>D. R. Congo | ♂ 55-89mm, ♀ 46-61mm | Subtribe Coryphocerina |

중앙아프리카에 서식한다. 이 아종은 흑갈색으로부터 녹색, 보라색까지 변이가 풍부하다. 머리방패는 마름모형이며 뿔은 하나로 길고 크게 발달하여 위쪽으로 굽어있으며 말단부에 돌기를 가지고 있다. 앞다리종아리마디에 돌기가 큰 톱날처럼 발달하였다.

♂

콩고민주공화국 산 69mm
(N. Klvu Lake, D. R. Congo. 2006. 6.)

♀

콩고민주공화국 산 58mm
(N. Klvu Lake, D. R. Congo. 2006. 8.)

# Flower Beetles / Tribe 5. Goliathini

This subspecies inhabits Central Africa. The cephalic shield of the species is lozenge-shaped and the single, long and big horn is bent upward with some protrusions at the tip. Some protrusions are well-developed like saw blades on the frontal calf joints(protibia).

[Distribution] Congo, D.R. Congo, Uganda, Luanda, Burundi

토르콰타-우간덴시쓰대왕귀신꽃무지 갈색형

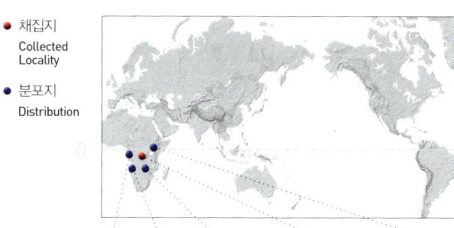

♀

콩고민주공화국 산 61mm
(N. Klvu Lake, D. R. Congo. 2007. 6.)

♂

콩고민주공화국 산 64mm
(N. Klvu Lake, D. R. Congo. 2007. 6.)

## 꽃무지아과 / Cetoniinae

**Mecynorhina**

| 학 명 Scientific name | 채집국 Collected locality | 크 기 Size | 참 고 Remark |
|---|---|---|---|
| 토르콰타-우간덴시쓰대왕귀신꽃무지<br>*Mecynorhina torquata ugandensis* (kraatz, 1907) |  콩고민주공화국<br>D. R. Congo | ♂ 64-76mm, ♀ 56-61mm | Subtribe Coryphocerina |

중앙아프리카에 서식한다. 이 아종은 *M. t. ugandensis*의 녹색형이다. 가슴부위가 광택이 없는 녹색을 띤다. 이 종의 딱지날개는 선명한 적자색을 띤다. 세로줄무늬의 변이가 다양하다. *M. t. ugandensis*의 공통된 특징으로 몸빛은 광택이 없고 다리 모두는 강한 광택을 지니고 있다.

♂

콩고민주공화국 산 76mm
(N. Klvu Lake, D. R. Congo. 2007. 6.)

♀

콩고민주공화국 산 56mm
(N. Klvu Lake, D. R. Congo. 2007. 6.)

# Flower Beetles / Tribe 5. Goliathini

This subspecies, inhabiting Central Africa, is the green type of *M. t. ugandensis*. Its thorax is green with no gloss while the elytra (hard wings) are reddish-purple and vivid. Its stripes show a wide range of variations. What is common for all *M. t. ugandensis* individuals is a body with no gloss and legs with some strong gloss.

[Distribution] Congo, D.R. Congo, Uganda, Luanda, Burundi

토르콰타-우간덴시쓰대왕귀신꽃무지 녹색형

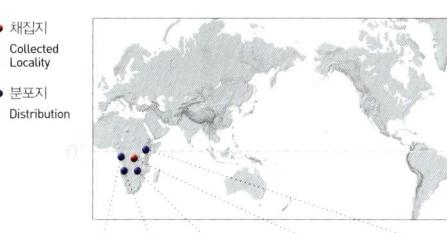

♀

콩고민주공화국 산 56mm
(N. Klvu Lake, D. R. Congo. 2006. 8.)

♂

콩고민주공화국 산 76mm
(N. Klvu Lake, D. R. Congo. 2006. 8.)

## 꽃무지아과 / Cetoniinae

**Megalorhina**

| 학 명 Scientific name | 채집국 Collected locality | 크 기 Size | 참 고 Remark |
|---|---|---|---|
| 하리시큰뿔꽃무지<br>*Megalorhina harrisi harrisihi* Westwood, 1847 | 우간다<br>Uganda | ♂ 28-45mm, ♀ 30-34mm | Subtribe Coryphocerina |

중앙아프리카에 서식한다. 이 종은 7~8아종으로 분류되고 있으며 몸빛과 머리뿔의 변이가 아종별로 차이가 있다. 이 아종의 몸빛은 광택이 없는 암녹색으로 가슴과 딱지날개에 황색의 테두리와 점무늬를 가지고 있다. 머리뿔은 머리방패와 함께 발달하였고 말단부는 두 갈래로 갈라져 위로 굽어있다. 다리에는 광택을 지닌다.

This species, which inhabits Central Africa, is divided into 7~8 subspecies, has a dark-green body with no gloss while some outer stripes and dot patterns are seen on the thorax and elytra (hard wings). The cephalic horn is developed with the cephalic shield, bent upward and divided into two at the tip. It has glossy legs.

[Distribution] D.R. Congo, Uganda, Ethiopia, Kenya

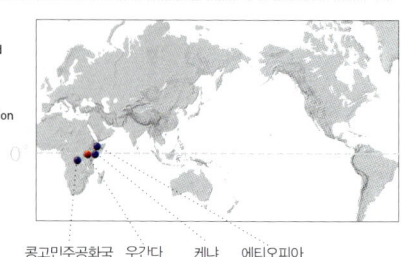

♂

우간다 산 40mm
(Kibale Forest Kabarole, W. Uganda. 2005. 11.)

♀

우간다 산 32mm
(Kibale Forest Kabarole, W. Uganda. 2005. 11.)

# Flower Beetles / Tribe 5. Goliathini

## Megalorhina

| 학 명 Scientific name | 채집국 Collected locality | 크 기 Size | 참 고 Remark |
|---|---|---|---|
| 하리시-엑스미아큰뿔꽃무지<br>*Megalorhina harrisi eximia* Aurivilus, 1886 | 카메룬<br>Cameroon | ♂ 34-55mm, ♀ 32-38mm | Subtribe Coryphocerina |

중앙아프리카에 서식한다. 이 종은 7~8아종을 가지고 있다. 이 아종들은 몸빛의 변이와 머리뿔의 변이가 개체별로 다르다. 그 중 *M. h. eximia*는 특히 몸빛의 변이가 다양하다. 이 종은 카메룬산으로 황색을 띠고 있다.

This species, which inhabits Central Africa, is divided into 7~8 subspecies, which show individual difference in their body colors and cephalic-horn shapes. Especially, *M. h. eximia* has a wide range of body-color variations. Native to Cameroon, this species has a yellow body.

[Distribution] Central Africa Rep., Cameroon, Congo

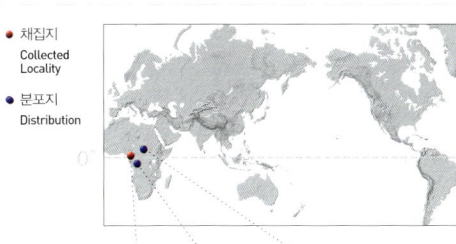

♀

카메룬 산 38mm
(Mt. Cameroun, Cameroon. 2005. 6.)

♂

카메룬 산 45mm
(Mt. Cameroun, Cameroon. 2005. 6.)

# 꽃무지아과 / Cetoniinae

## Mystroceros

| 학 명 Scientific name | 채집국 Collected locality | 크 기 Size | 참 고 Remark |
|---|---|---|---|
| 로우예리쌍뿔꽃무지<br>*Mystroceros rouyeri* Janson, 1907 | 인도네시아<br>Indonesia | ♂♀ 18.4–22.1mm | Subtribe Coryphocerina |

인도네시아 수마트라와 자바 섬에 서식한다. *Mystroceros*속에는 *M. macleayi*, *M. rouyeri* 두 종이 있다. 몸빛은 암, 수 모두 강한 광택의 검정색을 띠고 있으며 녹색, 적색, 보라색의 변이가 있다. 머리뿔은 기저부에서부터 두 갈래로 나뉘어져 발달하였으며 머리방패에 두 개의 작은 돌기가 있다.

This genus, consist of *M. macleayi* and *M. rouyeri*, inhabits Sumatra Island and Java Island, Indonesia. Both males and females have a black body with strong gloss while green, red and violet variations have been discovered. The cephalic horn is two-forked from the lower part and the cephalic shield has two small protrusions.

[Distribution]  Sumatera Island, Java Island

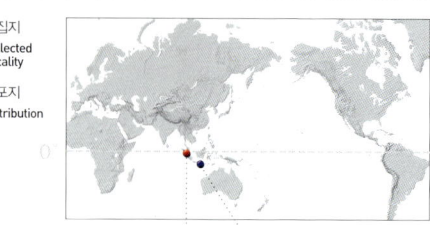

♂

수마트라 서부 산 19.5mm
(W. Sumatra, Indonesia. 2002. 6)

♀

수마트라 서부 산 20mm
(W. Sumatra, Indonesia. 2002. 6)

Flower Beetles / Tribe 5. Goliathini

## Neophaedimus

| 학 명 Scientific name | 채집국 Collected locality | 크 기 Size | 참 고 Remark |
|---|---|---|---|
| 아우조욱시뿔꽃무지<br>*Neophaedimus auzouxi* Lucas, 1870 | 중국<br>China | ♂25-32mm, ♀ 22-27mm | Subtribe Coryphocerina |

중국 대륙에 서식한다. 이종의 몸빛은 부드러운 광택의 적갈색으로 가슴양쪽에 한 쌍의 검은 반점과 두터운 세로줄을 가지고 있다. 배면은 전체가 검정색이다. 머리뿔은 머리방패와 함께 발달하여 말단부는 두 갈래로 갈라져 안쪽으로 굽어있으며 가슴뿔이 뾰족하게 발달하여 전방향으로 향하고 있다.

This species, which inhabits the mainland of China, has a softly-glossy and reddish-brown body with a pair of black speckles and thick stripes on both sides of the thorax while the back is entirely black. The cephalic horn, developed with the cephalic shield, is divided into two at the tip and bent inward while the pointed pronotum horn points forward.

[Distribution] China(Sichuan Prov. Yunnan Prov. Shaanxi Prov. Guansu Prov.)

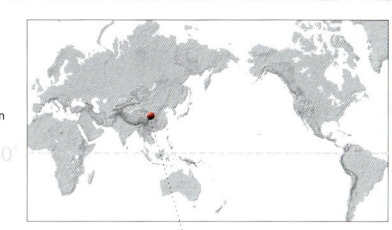

중국(사천성, 운남성, 산시성, 감서성)
China(Sichuan Prov., Yunnan Prov., Shaanxi Prov., Guansu Prov.)

중국 사천성 산 25mm
(Mt. Jinfo-shan, Sichuan, China. 2004. 6.)

## 꽃무지아과 / Cetoniinae

**Neptunides**

| 학 명 Scientific name | 채집국 Collected locality | 크 기 Size | 참 고 Remark |
|---|---|---|---|
| 스탠레이넵튠꽃무지<br>*Neptunides stanleyi* Bertoloni, 1855 |  콩고민주공화국<br>D. R. Congo | ♂ 29-36mm, ♀ 27-31mm | Subtribe Coryphocerina |

중앙아프리카에 서식한다. 이 종은 4아종이 있다. 몸빛은 강한 광택의 가슴과 딱지날개는 녹색/녹색, 적갈색/적황색, 녹색/황록색, 검정/갈색 등 변이가 다양하다. 이 종의 앞다리넓적다리와 딱지날개 가장자리는 적황색을 띤다. 머리뿔은 머리방패와 함께 발달되어 마치 왕관모양을 하고 있다.

Inhabiting Central Africa, this species is divided into four sub species. Its thorax and elytra (hard wings) have some strong gloss, showing diversified variations: green / green; reddish-brown / reddish-yellow; green / yellowish-green; black / brown, etc. The forelegs, thighs and the edge of the elytra (hard wings) are reddish-yellow while the cephalic horn, developed with the cephalic shield, resembles a crown.

[Distribution] Central Africa Rep., Cameroon, Gabon, D. R. Congo, Uganda, Kenya

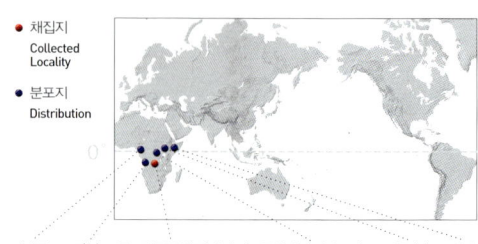

♂

콩고민주공화국 산 36mm
(Kivu, D. R. Congo. 2000. 7.)

♀

콩고민주공화국 산 31mm
(Kivu, D. R. Congo. 2000. 7.)

Flower Beetles / Tribe 5. Goliathini

## Neptunides

| 학 명 Scientific name | 채집국 Collected locality | 크 기 Size | 참 고 Remark |
|---|---|---|---|
| 폴리크로우스넵튠꽃무지<br>*Neptunides polychrous polychrous* Thomson, 1879 | 탄자니아<br>Tanzania | ♂ 23-32mm, ♀ 23-28mm | Subtribe Coryphocerina |

아프리카 탄자니아 동부에 서식한다. 이 종은 5아종이 있다. 몸빛은 강한 광택을 지니며 다양하다. 이 종은 머리와 딱지날개는 녹색, 가슴은 검정색을 띤다. 머리뿔은 머리방패와 함께 발달되어 마치 왕관모양을 하고 있다.

Inhabiting the eastern part of Tanzania, this species is divided into five subspecies. Its body, highly glossy, shows colorful variations. This species has a green and yellow body with the thorax speckled and patterned in dark brown. The cephalic horn, developed with the cephalic shield, resembles a crown.

[Distribution] Eastern Tanzania

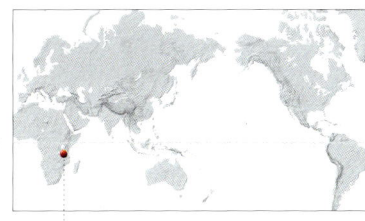

탄자니아 동부
E. Tanzania

♀

탄자니아 산 23mm
(Uluguru Mts., Tanzania. 2002. 4.)

♂

탄자니아 산 23mm
(Uluguru Mts., Tanzania. 2002. 4.)

## 꽃무지아과 / Cetoniinae

**Neptunides**

| 학 명 Scientific name | 채집국 Collected locality | 크 기 Size | 참 고 Remark |
|---|---|---|---|
| 폴리크로우스넵튠꽃무지<br>*Neptunides polychrous* Thomson, 1879 | 탄자니아 Tanzania | ♂ 27-32mm, ♀ 26-31mm | Subtribe Coryphocerina |

아프리카 탄자니아 동부에 서식한다. 이 종은 5아종이 있다. 몸빛은 강한 광택을 지니며 다양하다. 이 종의 몸빛은 녹색바탕에 황색을 지니며 가슴에 암갈색의 반점과 무늬를 가지고 있다. 머리뿔은 머리방패와 함께 발달되어 마치 왕관모양을 하고 있다.

Inhabiting the eastern part of Tanzania, this species is divided into five subspecies. Its body, highly glossy, shows colorful variations. This species has a green and yellow body with the thorax speckled and patterned in dark brown. The cephalic horn, developed with the cephalic shield, resembles a crown.

♂ 탄자니아 산 31mm (Uluguru Mts., Tanzania. 2002. 4.)

♀ 탄자니아 산 31mm (Uluguru Mts., Tanzania. 2002. 4.)

# Flower Beetles / Tribe 5. Goliathini

이 종은 머리와 딱지날개는 갈색, 가슴은 검정색을 띤다. 암컷은 머리와 가슴, 다리가 녹색을 띠며 가슴에 검정색 반점과 무늬를 가지고 있다.

This species is brown in the cephalic part and elytra (hard wings) and black in the thorax. The cephalic part, thorax and legs of females are green while some black speckles and patterns are seen on the thorax.

[Distribution] Eastern Tanzania

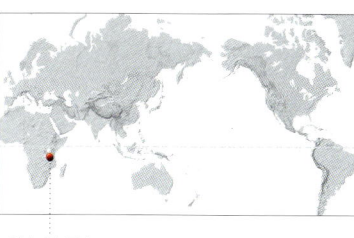

탄자니아 동부
E. Tanzania

♀

탄자니아 산 29mm
(Uluguru Mts., Tanzania. 2002. 4.)

♂

탄자니아 산 31mm
(Uluguru Mts., Tanzania. 2002. 4.)

## 꽃무지아과 / Cetoniinae

*Rhamphorrhina*

| 학 명 Scientific name | 채집국 Collected locality | 크 기 Size | 참 고 Remark |
|---|---|---|---|
| 베르토로니흰큰머리꽃무지<br>*Rhamphorrhina bertolonii* Lucas, 1879 | 탄자니아<br>Tanzania | ♂ 22-33mm, ♀ 23-27mm | Subtribe Coryphocerina |

아프리카, 탄자니아 동부에 서식한다. 이 종의 몸빛은 녹색, 밝은녹색, 적색 등으로 다양한 변이가 있다. 머리 중앙부와 가슴테두리, 딱지날개는 황백색이며 위와 아래 양쪽에 검은 반점을 가지고 있다. 머리뿔은 머리방패와 함께 발달되어 마치 왕관모양을 하고 있다. 앞발톱마디에 황색의 강모를 가지고 있다.

Inhabiting the eastern part of Tanzania, this species shows numerous body-color variations: green, bright green, red, etc. The cephalic center, edge of the thorax and elytra (hard wings) are yellowish-white while some black speckles are seen both at the top and bottom. The cephalic horn, developed with the cephalic shield, resembles a crown. The frontal tarsi are covered with yellow bristles.

[Distribution] Tanzania, Kenya

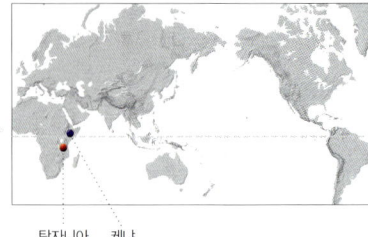

● 채집지 Collected Locality
● 분포지 Distribution

탄자니아 Tanzania  케냐 Kenya

♂

탄자니아 산 32mm
(Mt. Usambara, Tanzania. 2000. 10)

♀

탄자니아 산 27mm
(Mt. Usambara, Tanzania. 2000. 10)

# Flower Beetles / Tribe 5. Goliathini

## *Rhamphorrhina*

| 학 명 Scientific name | 채집국 Collected locality | 크 기 Size | 참 고 Remark |
|---|---|---|---|
| 스프레덴스흰큰머리꽃무지<br>*Rhamphorrhina splendens* Bertoloni, 1855 | 짐바브웨 Zimbabwe | ♂ 24-34mm, ♀ 22-26mm | Subtribe Coryphocerina |

남아프리카 중동부에 서식한다. 이 종은 *R. s. petersiana*와 함께 2아종이 있다. 몸빛은 녹색과 적색이 있으며 머리 중앙부와 가슴테두리, 딱지날개는 백색이며 세로줄무늬를 가지고 있다. 머리방패는 직사각형 처럼 넓적하게 발달하였다.

Inhabiting the mid-to-eastern parts of South Africa, this species has two subspecies with *R. s. petersiana*. Its body color is green or red while the cephalic center, pronotum edge and elytra (hard wings) are white and striped. The cephalic shield is developed in a wide rectangular shape.

[Distribution] Southern Zimbabwe, Northern Rep.of South Africa

♀

짐바브웨 산 22mm
(Guruve, Mashonaland, Zimbabwe. 2004. 1.)

♂

짐바브웨 산 33mm
(Guruve, Mashonaland, Zimbabwe. 2004. 1.)

# 꽃무지아과 / Cetoniinae

## Pseudotorynorrhina

| 학 명 Scientific name | 채집국 Collected locality | 크 기 Size | 참 고 Remark |
|---|---|---|---|
| 풍이<br>*Pseudotorynorrhina japonica* Hope, 1841 | 대한민국<br>Korea | ♂♀ 23.0-31.5mm | Subtribe Coryphocerina |

한반도를 비롯한 동아시아에 서식한다. *Pseudotorynorrhina*속은 3종이 포함되어 있다. 이 종의 몸빛은 광택을 지니며 서식지와 관계없이 적갈색, 녹색, 녹색/적갈색, 검정색 등 변이가 다양하다. 점각이 비교적 뚜렷하다.

This genus, which is divided into three subgroups, inhabits East Asia, including the Korean Peninsula. It has a glossy body and a wide range of body-color variations regardless of the habitat: dark brown; green; green / dark brown; black, etc. It has relatively prominent stipples.

[Distribution] Korea, Japan, China

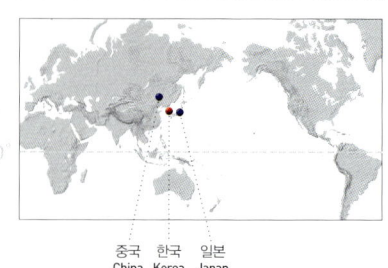

♀

전북 고창 산 25mm
(Mt. Seonunsan, Gochang-gun, Jeollabuk-do, S. Korea. 1990. 5.)

♀

전북 부안 산 24mm
(Mt. Byeonsan, Buan-gun, Jeollabuk-do, S. Korea. 1987. 5.)

Flower Beetles / Tribe 5. Goliathini

## Rhomborhina

| 학 명 Scientific name | 채집국 Collected locality | 크 기 Size | 참 고 Remark |
|---|---|---|---|
| 레스프렌덴스금광풍뎅이<br>*Rhomborhina resplendens* (Swartz, 1817) | 타이<br>Thailand | ♂♀ 27-40mm | Subtribe Coryphocerina |

인도 동북부, 인도차이나 북부, 중국남동부에 넓게 서식한다. 이 종의 몸빛은 금속성 광택이 강하다. 소순판 주변으로 딱지날개의 색이 청록색을 띤다. 아래의 사진처럼 모든 꽃무지는 딱지날개를 닫고 비행하는 특징을 가진다.

This species, whose body is highly metallic-glossy, has a wide habitat across the northeastern part of India, northern part of Indochina and southeastern part of China. Around the scutellum, the color of the elytra(hard wings) is greenish-blue. Like the picture below, all Cetoniid beetles fold their elytra(hard wings) while flying.

[Distribution] Northastern India, Nepal, Thailand, Laos, Vietnam, Southeastern China, Hainan Island

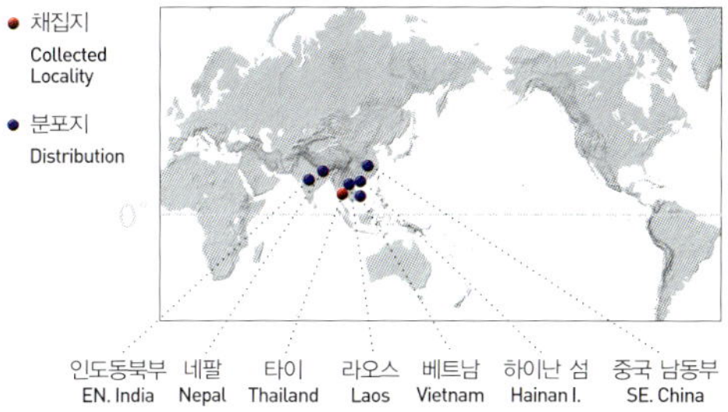

♀

타이 산 37mm
(Chiang-Rai, N. Thailand. 2001. 4.)

## 꽃무지아과 / Cetoniinae

**Stephanorrhina**

| 학 명 Scientific name | 채집국 Collected locality | 크 기 Size | 참 고 Remark |
|---|---|---|---|
| 구타타기린뿔꽃무지<br>*Stephanorrhina guttata* Olivier, 1789 | 카메룬<br>Cameroon | ♂ 22-29mm, ♀ 22-27mm | Subtribe Coryphocerina |

중앙아프리카 북서부에 서식한다. *Stephanorrhina*속은 10종이 알려져 있으며 *S. guttata*는 6아종으로 분류된다. 몸빛은 광택을 지니며 녹색바탕에 머리방패, 가슴테두리, 다리는 황적색을 띤다. 딱지날개는 흰 반점들이 있다. 특히 소순판과 딱지날개 중앙부는 적색이 강하다.

Inhabiting the northwestern part of Central Africa, the *Stephanorrhina* genus is known to be divided into 10 species while *S. guttata* is classified into 6 subspecies. It has a glossy and green body while the cephalic shield, thorax edge and legs are yellowish-red. Some white speckles are seen on the elytra (hard wings). It is especially red at the center of the elytra (hard wings) and the scutellum.

[Distribution] Senegal, Guinea , Cote Divoire, Burkina Faso, Benin, Cameroon

♂

카메룬 산 29mm
(Cameroon. 2002. 10.)

♀

카메룬 산 27mm
(Cameroon. 2002. 10.)

# Flower Beetles / Tribe 5. Goliathini

## Stephanorrhina

| 학 명 Scientific name | 채집국 Collected locality | 크 기 Size | 참 고 Remark |
|---|---|---|---|
| 줄리아기린뿔꽃무지<br>*Stephanorrhina julia* (Waterhouse, 1879) | 카메룬<br>Cameroon | ♂ 25-28mm, ♀ 23-27mm | Subtribe Coryphocerina |

아프리카 카메룬 서부에 서식한다. 몸빛은 광택을 지니며 녹색바탕에 머리방패, 가슴테두리, 다리는 *S. guttata*와 비교하여 황적색이 더 강하다. 딱지날개는 황백색 반점들이 있다. 특히 소순판과 딱지날개 중앙부는 적색이 강하다.

Inhabiting the western part of Cameroon, this species has a glossy and green body while the cephalic shield, thorax edge and legs are more yellowish-red than those of *S. g.aschantica*. Some yellowish-white speckles are seen on the elytra (hard wings). It is especially red at the center of the elytra (hard wings) and the scutellum.

[Distribution] Western Cameroon

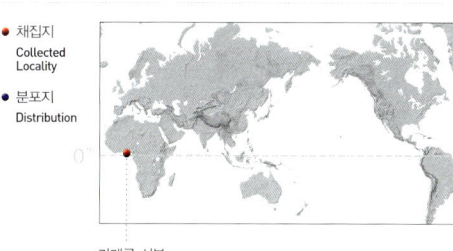

● 채집지 Collected Locality
● 분포지 Distribution

카메룬 서부
W. Cameroon

♀

카메룬 산 27mm
(Mt. Cameroun, Cameroon. 2002. 10.)

♂

카메룬 산 27mm
(Mt. Cameroun, Cameroon. 2002. 10.)

## 꽃무지아과 / Cetoniinae

**Taurrhina**

| 학 명 Scientific name | 채집국 Collected locality | 크 기 Size | 참 고 Remark |
|---|---|---|---|
| 롱기셉스앞뿔꽃무지<br>*Taurrhina longiceps* Kolbe, 1892 | 토고 Togo | ♂ 24-32mm, ♀ 23-27mm | Subtribe Coryphocerina |

중앙아프리카 중앙부에 서식한다. *Taurrhina*속은 3종이 알려져 있으며 *Taurrhina longiceps*종은 5아종이 알려져 있으며 도판은 원명아종이다. 몸빛은 광택을 지니며, 녹색바탕에 가장자리들은 청록색을 띠고 있다. 머리방패는 움푹 파여저있으며 기저부 중앙에 반타원형의 황색의 돌기가 솟아 있다. 복마디 중앙부에 적황색의 세로줄무늬가 있다.

Inhabiting the middle region of Central Africa, the *Taurrhina* genus is divided into three species with the *Taurrhina longiceps* species including five subspecies. Its body is glossy and green while the edge is greenish-blue. The cephalic shield is dented and some yellow, semi-oval protrusions are developed in the center of the lower part. The center of the abdominal sternites has some reddish-yellow stripes.

[Distribution] Togo, Nigeria, Central Africa Rep. , Cameroon, D.R. Congo, Congo

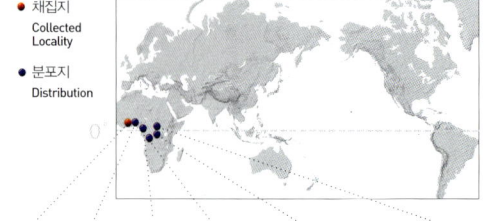

채집지 Collected Locality
분포지 Distribution

토고 Togo | 나이지리아 Nigeria | 카메룬 Cameroon | 콩고 Congo | 콩고민주공화국(자이르) D.R. Congo(Zaire) | 중앙아프리카공화국 Central Africa Rep.

♂

토고 산 31mm
(Kuma Forest, Kpalime, Togo. 2007. 9.)

## Flower Beetles / Tribe 5. Goliathini

### Pedinorrhina

| 학 명 Scientific name | 채집국 Collected locality | 크 기 Size | 참 고 Remark |
|---|---|---|---|
| 셉타노랑띠꽃무지<br>*Pedinorrhina septa* Harold, 1879 | 카메룬 Cameroon | ♂♀ 17-20mm | Subtribe Coryphocerina |

중앙아프리카에 넓게 서식한다. *Pedinorrhina*속은 8종이 알려져 있으며 *P. septa*는 두 개의 아종을 가지고 있다. 몸빛은 약한 광택의 검정색을 띤다. 딱지날개는 황색의 띠를 가지고 있다.

Having a wide habitat across Central Africa, the *Pedinorrhina* genus is divided into eight species with *P. septa* having two subspecies. Its body is slightly glossy and black while the elytra (hard wings) have some yellow stripes.

[Distribution] Central Africa Rep. , Cameroon, Guinea, D.R. Congo, Uganda

♀

카메룬 산 18mm
(Kupe, Cameroon. 2003. 10.)

♂

카메룬 산 19mm
(Kupe, Cameroon. 2003. 10.)

## 꽃무지아과 / Cetoniinae

**Tmesorrhina**

| 학 명 Scientific name | 채집국 Collected locality | 크 기 Size | 참 고 Remark |
|---|---|---|---|
| 알페스트리스꽃무지<br>*Tmesorrhina alpestris* Kolbe, 1892 | 카메룬 Cameroon | ♂♀ 23-31mm | Subtribe Coryphocerina |

중앙아프리카 서부에 서식한다. *Tmesorrhina*속은 소형종으로 14종이 알려져 있으며 녹색을 띠고 있다. *T. alpestris*종은 *T. a. alpestris*, *T. a. bafutensis* 두 아종으로 분류된다.

Inhabiting the western part of Central Africa, the *Tmesorrhina* genus, relatively small and green, is divided into 14 species. The *T. alpestris* species is divided into two subspecies of *T. a. alpestris*, *T. a. bafutensis*.

[Distribution] Cameroon

● 채집지 Collected Locality
● 분포지 Distribution

카메룬 Cameroon

♂

카메룬 산 29mm
(SW. Cameroon 2002. 10.)

## Flower Beetles / Tribe 5. Goliathini

### Tmesorrhina

| 학 명 Scientific name | 채집국 Collected locality | 크 기 Size | 참 고 Remark |
|---|---|---|---|
| 알페스트리-바후텐시스꽃무지<br>*Tmesorrhina alpestris bafutensis* F.Darge & P.Darge, 1988 | 카메룬<br>Cameroon | ♂♀ 22-30mm | Subtribe Coryphocerina |

중앙아프리카에 서부에 서식한다. 이 아종은 *T. a. alpestris*와 비교하여 딱지날개에 암녹색의 세로줄 무늬가 있다.

This subspecies, which inhabits the western part of Central Africa, has dark-green stripes on the elytra (hard wings) unlike *T. a. alpestris*.

[Distribution] Cameroon

♀

카메룬 산 29.2mm
(Mt. Cameroun, Cameroon, 2002. 10.)

## 꽃무지아과 / Cetoniinae

**Tmesorrhina**

| 학 명 Scientific name | 채집국 Collected locality | 크 기 Size | 참 고 Remark |
|---|---|---|---|
| 이리스긴몸광꽃무지<br>*Tmesorrhina iris* | 부룬디 Burundi | ♂♀ 23.6-29mm | Subtribe Coryphocerina |

아프리카 중부에서 서부에 서식한다. 이 종은 T. alpestris와 외형적 구분은 어려우나 보다 몸빛이 밝은 녹색이며 복면의 머리, 가슴, 넓적마디들이 검정색임이 두 종을 구분하는 결정적 요소들이 된다.

The green body color of this species, inhabiting the mid-to-western parts of Africa, is slightly brighter though very similar to that of *T. alpestris*. Its black cephalic part, thorax and femur distinguish this species from *T. a. alpestris*.

[Distribution] Gabon, Congo, D.R. Congo, Rwanda, Uganda, Burundi, Tanzania, Kenya

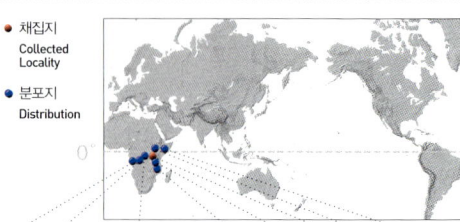

↑
부룬디 산 29mm
(Buhongo, Burundi. 2000. 1.)

Flower Beetles / Tribe 5. Goliathini

## Tmesorrhina

| 학 명 Scientific name | 채집국 Collected locality | 크 기 Size | 참 고 Remark |
|---|---|---|---|
| 라에타연보석꽃무지<br>*Tmesorrhina laeta* Moser, 1913 | 콩고민주공화국<br>D.R. Congo | ♂♀ 22-24mm | Subtribe Coryphocerina |

중앙아프리카 서부에 서식한다. 이 종은 *T. l. laeta*, *T. l. pilosipes* 두 아종이 있다. *T. l. pilosipes*는 녹색을 띠지만 이 원명아종은 딱지날개가 연분홍색을 띠고 있다.

Inhabiting the western part of Central Africa, this species is divided into two subspecies of *T. l. laeta* and *T. l. pilosipes*. While the latter is green, this subspecies has soft-pink elytra (hard wings).

[Distribution] Cameroon, D.R. Congo, Congo

● 채집지 Collected Locality
● 분포지 Distribution

카메룬 Cameroon　콩고 Congo　콩고민주공화국(자이르) D.R. Congo(Zaire)

♂

콩고민주공화국 산 22mm
(Bandundu, Kikos, D.R.Congo. 2001. 2.)

## 꽃무지아과 / Cetoniinae

### Trigonophorus

| 학 명 Scientific name | 채집국 Collected locality | 크 기 Size | 참 고 Remark |
|---|---|---|---|
| 로칠디주걱턱꽃무지<br>*Trigonophorus rothschildi* Fairmaire, 1891 | 타이완<br>Taiwan | ♂♀ 28.7-34mm | Subtribe Coryphocerina |

중국 남부, 타이완에 서식한다. 아종 varians가 알려져 있다. *rothschildi*의 몸빛은 녹색, 녹갈색을 띠며, *varians*은 녹색, 황적색, 청색, 보라색을 띤다. 녹색 빛의 이 종은 마치 북극의 에스키모가 고래를 썰때 사용하는 반달형 칼처럼 생긴 뿔을 지니고 있다. 도판은 타이완산 아종인 *varians*이다.

This species, inhabiting the southern part of China and Taiwan, is divided into two subspecies of *rothschildi* and *varians*. The body color of the former is green or greenish-brown while that of the latter is green, yellowish-red, blue or violet. The cephalic horn of this green species looks like the crescent knife of the Inuits used when chopping a whale.

[Distribution] China, Taiwan

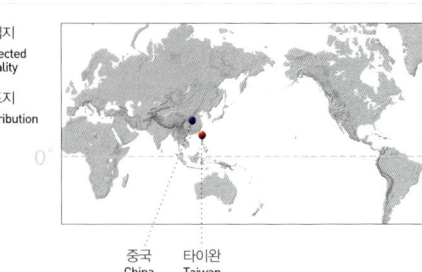

♂

타이완 산 37mm
(Taiwan. 2004. 11.)

♀

타이완 산 34mm
(Taiwan. 2004. 11.)

# Flower Beetles / Tribe 5. Goliathini

## Torynorrhina

| 학 명 Scientific name | 채집국 Collected locality | 크 기 Size | 참 고 Remark |
|---|---|---|---|
| 후라메아풍이<br>*Torynorrhina flammea flammea* Gestro, 1888 | 타이<br>Thailand | ♂♀ 29.5-36.5mm | Subtribe Coryphocerina |

인도 북부에서, 인도차이나 북부, 말레이반도, 중국 시후안성까지 서식한다. *Torynorrhina*속은 10종이 알려져 있다. *T. flammea*는 *T. chicheryi*와 더불어 두 아종으로 분류된다. 이 원아종은 녹색바탕의 적색형과 청색형이 있다. 다리와 복면은 청색이며 복마디는 검정색이다. 가슴복판 중앙부는 녹색을 띤다.

Inhabiting the northern part of India, northern part of Indochina, the Malay Peninsula and Sihuan Province, China, the *Torynorrhina* genus is divided into ten species. *T. flammea* and *T. chicheryi* constitute two subspecies groups. This species is classified into green and blue variations, both of which have the body color of green. The legs and back are blue while the abdominal sternites are black. The mesosternum is green.

[Distribution] India, Laos, Myanmar, Malaysia, Thailand, China

♀
적색형(Red Color)

타이 산 33mm
(Chiang Rai, N. Thailand. 2002. 10.)

## 꽃무지아과 / Cetoniinae

**Torynorrhina**

| 학 명 Scientific name | 채집국 Collected locality | 크 기 Size | 참 고 Remark |
|---|---|---|---|
| 후라메아풍이<br>*Torynorrhina flammea flammea* Gestro, 1888 | 타이<br>Thailand | ♂♀ 29.5-36.5mm | Subtribe Coryphocerina |

이 종은 원아종으로 청색형과 녹황색형이다.

♂
청색형(Blue Color)

타이 북부 산 34mm
(Chiang - Mai, N. Thailand. 2007. 9.)

♀
청색형(Blue Color)

타이 북부 산 34mm
(Chiang - Mai, N. Thailand. 2007. 9.)

# Flower Beetles / Tribe 5. Goliathini

This subspecies classified into blue and greenish-yellow variations.

[Distribution] India, Laos, Myanmar, Malaysia, Thailand, China(Yunnan)

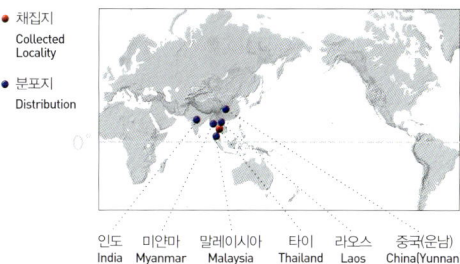

↑
녹황색형(Greenish Yellow Color)

타이 북부 산 29mm
(Chiang - Rai, N. Thailand. 2007.10.)

## 꽃무지아과 / Cetoniinae

**Torynorrhina**

| 학 명 Scientific name | 채집국 Collected locality | 크 기 Size | 참 고 Remark |
|---|---|---|---|
| 후라메아-키케리풍이<br>*Torynorrhina flammea chicheryi* Gestro, 1888 | 말레이시아 Malaysia | ♂♀ 29-37mm | Subtribe Coryphocerina |

말레이 반도에 서식한다. 이 아종은 청녹색으로 희귀하게 청색이 강한 개체도 발견된다.

♂

서말레이시아 산 35mm
(Cameton Highland, W. Malaysia. 2008. 5.)

## Flower Beetles / Tribe 5. Goliathini

Inhabiting the Malay Peninsula. This subspecies is classified into green and blue variations.

[Distribution] Malay Peninsula

말레이 반도
Malay Pe.

♀

서말레이시아 산 33mm
(Cameton Highland, W. Malaysia. 2008. 5.)

## 6. 꽃무지족(族)

이 족은 110속 1,050종에 이르는 거대한 족으로 전 세계의 꽃무지아과 중 30%를 차지한다. 한국을 포함한 구북구(Paleorctic region)에 서식하는 꽃무지는 주로 이 족에 속하며 종 수는 동양 열대구(Oriental region)와 에티오피아구(Ethiopian region)에 많다.
이 장에서는 5속(genus) 19종(species) 29개체를 수록하였다.

Consisting of 1,050 species under 110 genera, this large tribe amounts to 30% of all Cetoniidae of the world. Those Cetoniid beetles inhabiting the Paleorctic region, including the Korean Peninsula, fall into this tribe. The Oriental region and Ethiopian region are the areas with the biggest number of species in this tribe.
29 individuals of 5 genus, 19 species are included in this section.

Section 6
# Cetoniini

# 꽃무지아과 / Cetoniinae

## Glycyphana

| 학 명 Scientific name | 채집국 Collected locality | 크 기 Size | 참 고 Remark |
|---|---|---|---|
| 네점박이붉은띠꽃무지<br>*Glycyphana* sp. | 필리핀<br>Philippines | 17.5-18mm | |

*Glycyphana*속은 동남아시아와 아시아에 넓게 서식한다. 이 속은 6아과 80여종이 알려져 있다. 이 종은 필리핀에서 발견된 종명이 불확실한 종이다. 몸빛은 무광의 검정바탕에 흰점을 가지고 있으며 가슴의 띠와, 소순판이 적색인 것이 특징이다.

The *Glycyphana* genus lives throughout Asia, including Southeast Asia. Currently, 3 subfamilies and some 80 species are known to fall under this genus. The name of this species, discovered in the Philippines, has yet been established. It has white spots on its flat black body, and it is characterized by some red stripes on the thorax and the scutellum of the same color.

[Distribution] Philippines

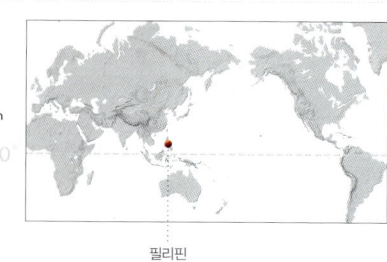

필리핀 산 18mm
(Philippines. 2008.)

필리핀 산 17.5mm
(Philippines. 2008.)

Flower Beetles / Tribe 6. Cetoniini

## Pachnoda

| 학 명 Scientific name | 채집국 Collected locality | 크 기 Size | 참 고 Remark |
|---|---|---|---|
| 에피피아타-활케이주홍테꽃무지<br>*Pachnoda ephippiata falkei* Rigout, 1989 |  르완다<br>Rwanda | ♂♀ 21.3-29mm | |

중앙아프리카 동부에 서식한다. 이 종은 *P.e.falkei*와 *P.e.francoisi* 두 개의 아종으로 나뉘어져 있다. 이 종은 황색바탕에 마치 로켓와 같은 검정색의 무늬를 가지는 큰 특징이 있다. 무늬는 개체별로 변이가 있어 이들을 구분하는 데에는 어려움이 있으나 배면의 경우 *P.e.falkei*는 황색, *P.e.francoisi*는 암갈색을 띠고 있어 이 두 아종을 분류하는 결정적인 요소가 된다.

This species, which is divided into two subspecies of *P.e.falkei* and *P.e.francoisi*, inhabits the eastern region of central Africa. They are characterized by some rocket-like black patterns on their yellow bodies. Though the pattern on the back, showing some individual variations, are not enough to divide the two, that of the former is yellow while that of the latter is dark brown, which facilitates their grouping.

[Distribution] Uganda, Rwanda, Kenya

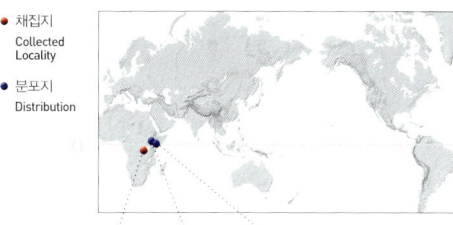

● 채집지 Collected Locality
● 분포지 Distribution

르완다 Rwanda  우간다 Uganda  케냐 Kenya

♂
르완다 산 29mm
(Save, Rwanda. 2002. 5.)

# 꽃무지아과 / Cetoniinae

## Pachnoda

**학 명 Scientific name**
에피피아타-환코이시주홍테꽃무지
*Pachnoda ephippiata francoisi* (Fabricus, 1775)

**채집국 Collected locality** 탄자니아 Tanzania

**크 기 Size** ♂♀ 25.3-26mm

**참 고 Remark**

중앙아프리카 서부에서 동부까지 서식한다. 색깔과 무늬도 비슷한 같은 종이지만 배면의 경우 *P.e.falkei*는 황색, *P.e.francoisi*는 암갈색으로 더 긴 털을 가지고 있어 두 종을 분류하는 결정적인 요소가 된다. 두 종 모두 배면에만 광택을 지닌다.

They inhabit the western to eastern region of central Africa. Though the two have similar body colors and patterns, *P.e.falkei* has a yellow back whereas *P.e.francoisi* has a dark brown one with some long hair, which facilitates their grouping. Both of them are glossy only in the back.

[Distribution] Togo, Tanzania

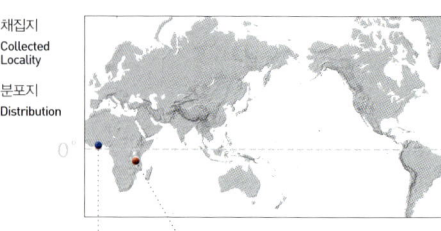
- 채집지 Collected Locality
- 분포지 Distribution

토고 Togo / 탄자니아 Tanzania

♂

탄자니아 산 26mm
(Mt. Usambara, Tanzania)

Flower Beetles / Tribe 6. Cetoniini

| 학 명 Scientific name | 채집국 Collected locality | 크 기 Size | 참 고 Remark |
|---|---|---|---|
| 마르기나타주홍테꽃무지<br>*Pachnoda marginata* Drury, 1773 | 코트디브아르<br>Cote divoire | ♂♀ 24.9-26.6mm | |

Pachnoda

중앙아프리카 서북부에 서식한다. 이 종은 *P. ephippiata* 비교하여 검정색의 무늬가 더욱 큰 타원형이며 배면은 전체가 검정색으로 털을 가지고 있지 않다.

This species inhabits the northwestern area of central Africa. Compared with *P.e.falkei* and *P.e.francoisi*, it has bigger and oval black patterns. Its back, all black, has no hair.

[Distribution] Guinea, Sierra Leone, Cote divoire, Togo, Nigeria

● 채집지 Collected Locality
● 분포지 Distribution

기니 Guinea | 시에라리온 Sierra Leone | 토고 Togo | 코트디브아르 Cote divoire(Ivory Coast) | 나이지리아 Nigeria

♂
코트디브아르 산 26mm
(Cote divoire. 2006.)

123

# 꽃무지아과 / Cetoniinae

## Protaetia

**학 명 Scientific name:** 쿠프레아점박이꽃무지 *Protaetia cuprea* (Fabricius, 1775)

**채집국 Collected locality:** 슬로바키아 Slovakia

**크 기 Size:** ♂♀ 18.6-26.3mm

**참 고 Remark:**

유럽, 중국, 사할린에 이르기까지 구북구(Paleorctic region) 전반에 넓게 서식한다. 이 종은 아속 *Potosia*에 속하며 구북구에 약 40종이 알려져 있다. 이 종의 몸빛은 황녹색, 금녹색, 녹색, 적동색, 녹황색 등 색체의 변이가 많고, 점무늬가 있는 것과 없는 것 등 서로 다른 종이라 할 만큼 변이가 많은 종이다. 모두가 광택을 지니고 있다.

This species has an expansive habitat in the Paleoarctic region from Europe and China to Sakhalin. Its body color shows a very wide range of individual variations from yellowish-green, golden-green, green, red-copper to greenish-yellow. So does its spotted pattern, some individuals having one unlike others. All of them are glossy.

[Distribution] Europe (Itallia, Slovakia, Turkey, Azerbaidzhan, Hungary, etc.), Sakhalin Island, Mongol, China

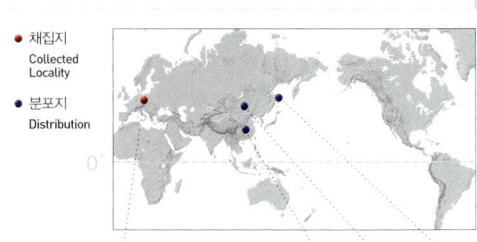

● 채집지 Collected Locality
● 분포지 Distribution

유럽(이탈리아, 터키, 아제르바이잔, 헝가리, 등)
Europe (Itallia, Slovakia, Turkey, Azerbaidzhan, Hungary, etc.)  중국 China  몽골 Mongol  사할린 Sakhalin I.

♀

슬로바키아 산 22mm
(Slovakia. 2004. 8)

Flower Beetles / Tribe 6. Cetoniini

## Proteatia

| 학 명 Scientific name | 채집국 Collected locality | 크 기 Size | 참 고 Remark |
|---|---|---|---|
| 쿠프레아—올리바세아꽃무지<br>*Proteatia cuprea olivacea* (Mulsant, 1842) | 프랑스 France | ♀ 20mm | |

유럽과 아시아에 서식한다. *Proteatia*속의 종수는 약 373여종이 알려져 있으며 이 종은 몸빛 전체가 녹갈색의 금속성 광택을 띠고 있으나 광택은 강하지 않다. 배면은 전체가 적색에 가까운 보라색을 띤다. 이 종은 *Potosia*로 분류되어 졌다.

The *Proteatia* genus, inhabiting Europe and Asia, is known to be divided into about 373 species. This species has a greenish-brown body with subtle metallic gloss. Its entire back is purple, close to red. This species used to fall under the genus of *Potosia*.

[Distribution] Europe(France)

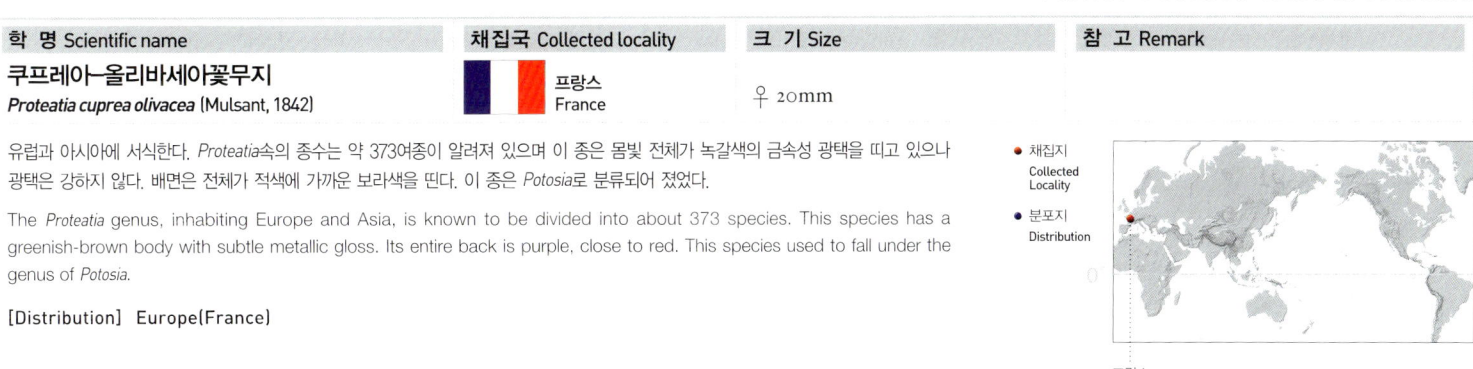

♀

프랑스 산 20mm
(Les Mayons, France. 1994. 6.)

## 꽃무지아과 / Cetoniinae

**Protaetia**

| 학 명 Scientific name | 채집국 Collected locality | 크 기 Size | 참 고 Remark |
|---|---|---|---|
| 쎄레비카점박이꽃무지<br>*Protaetia celebica* Wallace, 1867 | 인도네시아<br>Indonesia | ♂ ♀ 18-27mm | |

인도네시아 술라웨시 섬에 서식한다. 이 종의 몸빛은 검정색, 적동색, 황녹색 등 색체의 변이와 점무늬의 변이가 많다. 또한 가슴양쪽 둘레에 황색의 띠를 가지고 있는 것도 있다. 암컷은 수컷에 비하여 광택이 강하다.

This species inhabits Sulawesi Island, Indonesia. The body color and spotted pattern of this species show a wide range of individual variations from black, red-copper to greenish-yellow. Some individuals have yellow stripes along their thorax boundary. Females are glossier than their male counterparts.

[Distribution] Sulawesi Island

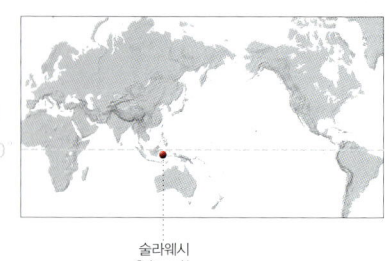

♀
인도네시아 산 24mm
(Palolo, Sulawesi, Indonesia. 2003. 7.)

♀
인도네시아 산 23mm
(Palolo, Sulawesi, Indonesia. 2003. 7.)

# Flower Beetles / Tribe 6. Cetoniini

## Protaetia

| 학 명 Scientific name | 채집국 Collected locality | 크 기 Size | 참 고 Remark |
|---|---|---|---|
| 녹스점박이큰꽃무지<br>*Protaetia nox* Janson, 1881 |  필리핀 Philippines | ♂♀ 23-30mm | |

필리핀의 섬에 서식한다. 이 종의 몸빛은 광택이 없는 검정색으로 벨벳과 같은 매우 짧은 미모로 덮여있다. 가슴뒤쪽으로 황백색의 띠를 가지고 있는 것도 있다. 딱지날개의 황백색 무늬는 변이가 많다.

Inhabiting the Philippines, this species has a flat black body covered with some short, fine hair like velvet. Some individuals have yellowish-white stripes on the back of the thorax. The yellowish-white pattern on the elytra (hard wings) shows diverse individual variations.

[Distribution] Philippines (Luzon Island, Mindoro Island, Leyte Island, Mindanao Island)

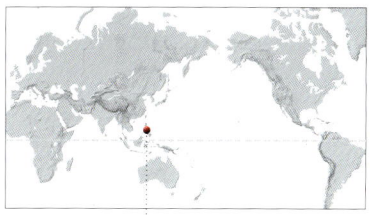

필리핀(루손 섬, 민도루 섬, 레이테 섬, 민다나오 섬)
Philippines(Luzon I., Mindoro I, Leyte I., Mindanao I.)

♀
필리핀 민다나오 산 28mm
(Mt. Apo, Mindanao I., Philippines 2002. 6.)

♂
필리핀 민다나오 산 27mm
(Mt. Apo, Mindanao I., Philippines 2002. 6.)

Protaetia

| 학 명 Scientific name | 채집국 Collected locality | 크 기 Size | 참 고 Remark |
|---|---|---|---|
| 루마위기점박이꽃무지<br>*Protaetia lumawigi* Arnaud, 1987 | 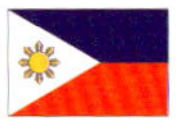 필리핀<br>Philippines | ♂♀ 20-28mm | |

필리핀에 서식한다. 이 종의 몸빛은 검정색, 적동색, 황녹색 등 색체의 변이와 점무늬의 변이가 많다. 또한 가슴양쪽 둘레에 황색의 띠를 가지고 있는 것도 있다. 암컷은 수컷에 비하여 광택이 강하다.

This species inhabits the Philippines. The body color and spot pattern of this species show a wide range of individual variations from black, red-copper to greenish-yellow. Some individuals have yellow stripes along their thorax boundary. Females are glossier than their male counterparts.

[Distribution] Philippines (Marinduque Island, Mindoro Island)

● 채집지 Collected Locality

● 분포지 Distribution

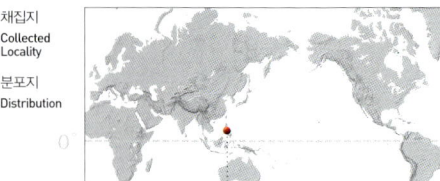

필리핀(마린두께 섬, 민도르 섬)
Philippines(Marinduque I., Mindoro I.)

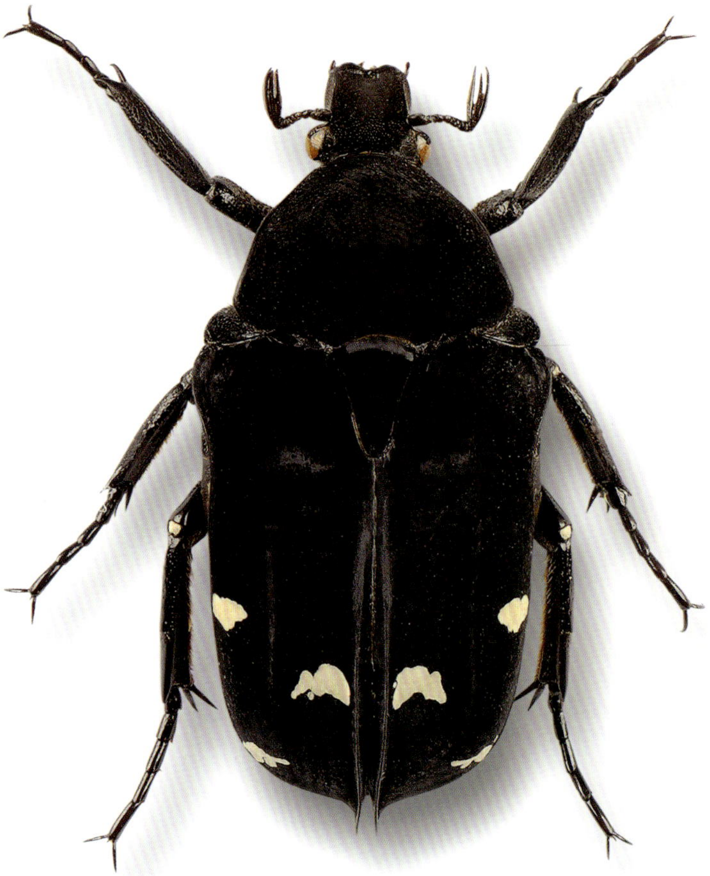

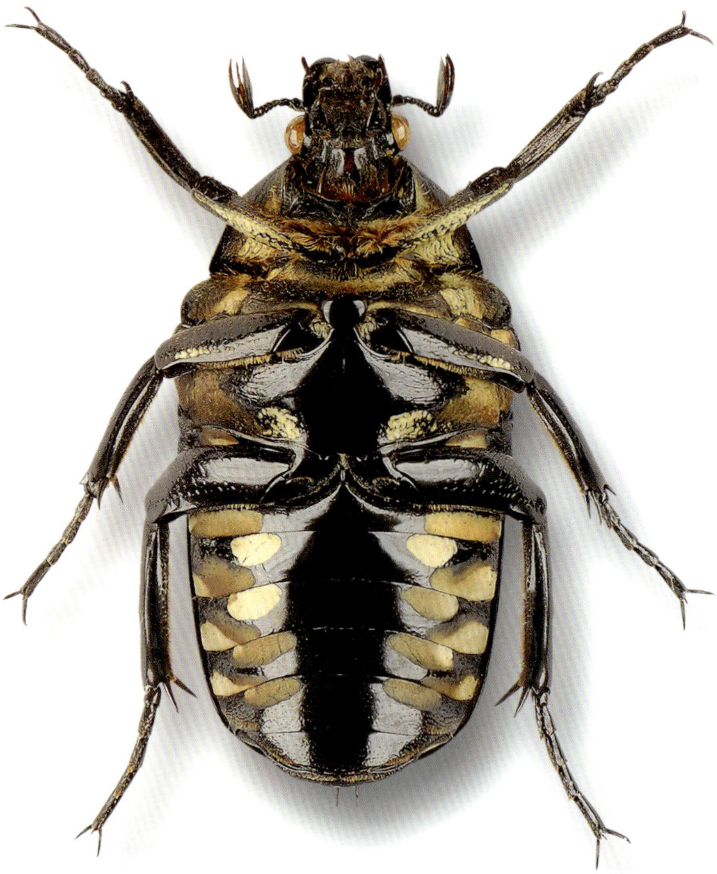

♂

필리핀 민다나오 산 28mm
(Mt. Halcon, Mindoro I., Philippines 2003. 8.)

# Flower Beetles / Tribe 6. Cetoniini

## Protaetia

| 학 명 Scientific name | 채집국 Collected locality | 크 기 Size | 참 고 Remark |
|---|---|---|---|
| 헝가리카꽃무지<br>*Protaetia hungarica* (Herbst, 1790) | 우크라이나<br>Ukraine | ♂♀ 21mm | |

동유럽에 서식한다. *Potosia*아속에 속하여, 이 속에는 약 40여종이 알려져 있다. 이 종은 녹색의 금속성 광택을 지니고 있으며 가슴에 네 개의 금색 점과 딱지날개에 반점들을 가지고 있지만 개체별로 변이가 있다.

The *Potosia* genus, which inhabits Eastern Europe, is divided into some 40 species. This species, having a metallic-green gloss, has four golden spots on its thorax and some speckles on its elytra (hard wings), though there are some individual variations.

[Distribution] Ukraina

♀

우크라이나 산 21mm
(Ukraine. 2004.)

## 꽃무지아과 / Cetoniinae

**Protaetia**

| 학 명 Scientific name | 채집국 Collected locality | 크 기 Size | 참 고 Remark |
|---|---|---|---|
| 괌꽃무지<br>*Protaetia guam* Arnaud, 1987 |  괌(미국령)<br>Guam(American territory) | ♂♀ 22-24mm | |

마크로네시아 위쪽, 괌 섬에 서식한다. 이 종의 몸빛은 녹동색의 금속광택이 강하다. 몸 전체의 점각이 두드러져 보인다. 복면 또한 녹동색으로 금속성광택이 강하며 다리에 털이 나있다.

This species inhabits the upper region of Micronesia and Guam. It has a copper-green, metallic-glossy body. The stipples throughout its body look prominent. Its back has the same copper-green color and the legs are hairy.

[Distribution] Guam Island

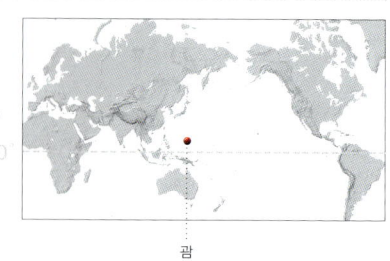

● 채집지 Collected Locality
● 분포지 Distribution

괌 Guam I.

♀

괌 산 24mm
(Guam. 2004.)

Flower Beetles / Tribe 6. Cetoniini

## Protaetia

| 학 명 Scientific name | 채집국 Collected locality | 크 기 Size | 참 고 Remark |
|---|---|---|---|
| 오리엔탈리스점박이꽃무지<br>*Protaetia orientalis* Gory & Percheron, 1833 | 타이완 Taiwan | ♂♀ 20-30.4mm | |

중국 남부, 타이완에 서식한다. 원아종과 더불어 *sakaii, submarmorea, tokarana* 4아종이 있다. 몸빛은 갈동색의 금속광택을 지닌다. 가슴과 딱지날개의 황갈색무늬는 일정한 규칙을 가지고 있으나 변이가 많다.

This species, inhabiting southern China and Taiwan, is usually grouped together with the 3 subspecies of *P.sakaii*, *P.submarmorea* and *P.tokarana*. It has a metallic-glossy, copper-brown body. The yellowish-brown pattern on the thorax and elytra (hard wings), though having regularity, shows a wide range of variations.

[Distribution] China, Taiwan

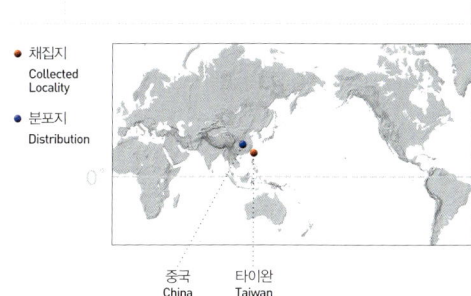

♀

타이완 산 20mm
(Taiwan. 2004. 7.)

♀

타이완 산 21mm
(Taiwan. 2004. 7.)

# 꽃무지아과 / Cetoniinae

**Protaetia**

| 학 명 Scientific name | 채집국 Collected locality | 크 기 Size | 참 고 Remark |
|---|---|---|---|
| 흰점박이꽃무지<br>*Protaetia brevitarsis seulensis* (Kolbe, 1879) | 대한민국<br>Korea | ♂♀ 17-23mm | |

한반도, 동시베리아, 중국, 일본에 서식한다. 이 종의 몸빛은 적동색, 흑동색으로 금속성 광택이 있으며 복면은 더욱 강하다. 몸 전체에 점각이 있으며 딱지날개의 미색 점무늬는 변이가 많다.

This species inhabits the Korean Peninsula, Eastern Siberia, China and Japan. Its body is copper-brown or copper-black with some metallic gloss, which looks stronger on the back. It has some stipples throughout its body. The pale yellow spot pattern on the elytra (hard wings) shows a wide range of variations.

[Distribution] Korea, China, Eastern Siberia, Japan

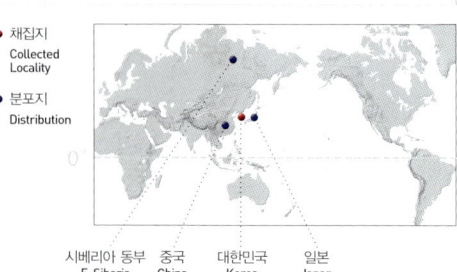

● 채집지 Collected Locality
● 분포지 Distribution

시베리아 동부 / E. Siberia  중국 / China  대한민국 / Korea  일본 / Japan

♂

전북 부안 산 23mm
(Mt. Byeonsan, Buan-gun, Jeollabuk-do, S. Korea. 2006. 5.)

Flower Beetles / Tribe 6. Cetoniini

## Protaetia

| 학 명 Scientific name | 채집국 Collected locality | 크 기 Size | 참 고 Remark |
|---|---|---|---|

**점박이꽃무지**
*Protaetia orientalis submarmorea* (Burmeister, 1842)

대한민국 / Korea

♂♀ 16-27mm

한반도를 비롯하여 인도, 중국, 일본에 이르기까지 넓게 서식한다. 이 종의 몸빛은 황동색, 녹색으로 금속성 광택이 있으며 복면은 더욱 강하다. *P. brevitarsis seulensis*에 비하여 몸 전체의 점각이 두드러져 있으며 복마디 양쪽으로 황백색의 반점이 있다. 딱지날개의 황백색의 점무늬는 변이가 많다.

This species has an expansive habitat from the Korean Peninsula, India, China to Japan. Its body is copper-brown or copper-green with some metallic gloss, which is more prominent on the back. Compared with *P. brevitarsis seulensis*, it shows more prominent stipples across the body, having some yellowish-white speckles on both sides of its abdominal segments. The yellowish-white spot pattern on the elytra (hard wings) shows a wide range of variations.

[Distribution] Korea, India, Himalaya Mountains, China, Taiwan, Guam, Japan

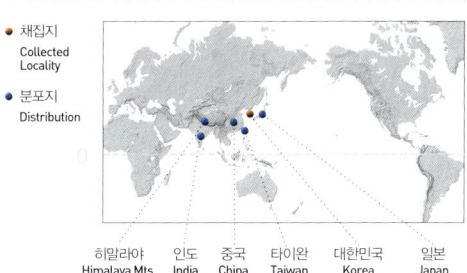

전북 부안 산 22mm
(Mt. Byeonsan, Buan-gun, Jeollabuk-do, S. Korea. 2006. 5.)

## 꽃무지아과 / Cetoniinae

**Protaetia**

| 학 명 Scientific name | 채집국 Collected locality | 크 기 Size | 참 고 Remark |
|---|---|---|---|
| 필리핀점박이꽃무지<br>*Protaetia phillppensis* (Fabricius, 1775) | 필리핀 Philippines | ♂♀ 16-21mm | |

필리핀에 서식한다. 이 종의 몸빛은 녹색, 암녹색의 금속성 광택이 있으며 가슴에는 한 쌍의 황백색 점무늬가 있고 양쪽으로 한 쌍의 띠를 가지고 있다. 딱지날개의 점무늬는 일정한 규칙을 가지고 있으나 변이가 많다.

This species, inhabiting the Philippines, has a green or dark green body with some metallic gloss. On this thorax is a pair of yellowish-white spot patterns with some bidirectional stripes. The spot pattern on the elytra (hard wings) shows a wide range of variations.

[Distribution] Philippines (Luzon Island, Mindoro Island)

필리핀(루손 섬, 민도루 섬)
Philippines(Luzon I., Mindoro I.)

♂

필리핀 루손 산 21mm
(Luzon I., Philippines. 2006. 6.)

♂

필리핀 루손 산 20mm
(Luzon I., Philippines. 2006. 6.)

Flower Beetles / Tribe 6. Cetoniini

## Protaetia

| 학 명 Scientific name | 채집국 Collected locality | 크 기 Size | 참 고 Remark |
|---|---|---|---|

스켑시아점박이꽃무지
*Protaetia scepsia* (Dohrn, 1872)

필리핀
Philippines

♂♀ 17-23mm

필리핀과 베트남에 서식한다. 이 종의 몸빛은 녹색, 암녹색의 금속성 광택이 있으며 가슴에는 한 쌍의 황백색 점무늬가 있고 양쪽으로 한 쌍의 띠를 가지고 있다. 딱지날개의 점무늬는 일정한 규칙을 가지고 있으나 변이가 많다.

This species, inhabiting the Philippines and Vietnam, has a green or dark green body with some metallic gloss. On this thorax is a pair of yellowish-white spot patterns with some bidirectional stripes. The spot pattern on the elytra (hard wings) shows a wide range of variations.

[Distribution] Philippines (Luzon, Camiguin, Samar, Leyte, Mindanao, Palawan Island ), Vietnam

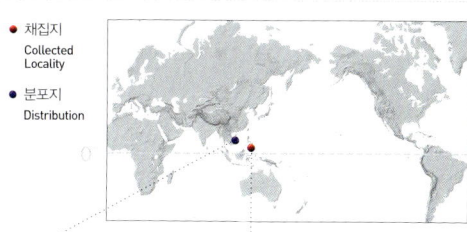

베트남 Vietnam
필리핀(루손섬, 카미구인섬, 사마르섬, 레이테섬, 민다나오섬, 팔라완섬)
Philippines (Luzon I., Camiguin I., Samar I., Leyte I., Mindanao I., Palawan I. )

♀

필리핀 민다나오 산 20mm
(Mt. Apo, Mindanao I., Philippines. 2007. 7.)

♀

필리핀 레이테 산 22mm
(Leyte I., Philippines. 2007. 7.)

## 꽃무지아과 / Cetoniinae

**Protaetia**

| 학 명 Scientific name | 채집국 Collected locality | 크 기 Size | 참 고 Remark |
|---|---|---|---|
| 우후리기큰점박이꽃무지<br>*Protaetia uhligi* Arnaud, 1987 |  필리핀<br>Philippines | ♂♀ 25-33mm | |

필리핀 루손 섬에 서식한다. 이 종의 몸빛은 적갈색으로 딱지날개는 더 검다. 딱지날개는 광택이 없으며 황백색의 큰 점무늬를 가지고 있으며 복면의 가슴과 복마디에 황백색의 무늬와 광택을 지니고 있다.

This species inhabits Luzon Island, Philippines. This species is dark red while its elytra (hard wings) are closer to black. The flat elytra (hard wings) have some sizable, yellowish-white spot pattern. On the glossy thorax and abdominal segment is a similar pattern.

[Distribution] Philippines(Luzon Island)

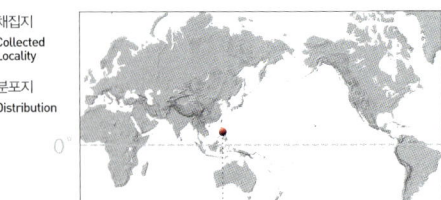

필리핀(루손 섬)
Philippines(Luzon I.)

↑
필리핀 루손 산 27mm
(Aurora Prov., Luzon I., Philippines. 2006. 6.)

Flower Beetles / Tribe 6. Cetoniini

## Protaetia

| 학 명 Scientific name | 채집국 Collected locality | 크 기 Size | 참 고 Remark |
|---|---|---|---|
| 베네라빌리스큰풀색꽃무지<br>*Protaetia venerabilis* Mohnike, 1873 | 필리핀 Philippines | ♂♀ 17-23mm | |

필리핀 루손 섬에 서식한다. 이 종의 몸빛은 광택이 없는 녹색을 띠고 있다. 가슴 양쪽으로 백색의 띠를 가지고 있으며 한 쌍의 매우 작은 점이 있다. 딱지 날개는 일정한 규칙의 점무늬를 가지고 있으나 변이가 많다.

This species, inhabiting Luzon Island, Philippines, has a flat green body. Some white stripes and a couple of minute spots are laterally present on the thorax. The elytra (hard wings), though having some regular spot pattern, show a wide range of variations.

[Distribution] Philippines(Luzon, Polillo, Catanduanes, Marinduque, Mindoro Island)

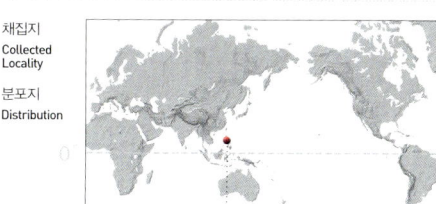

필리핀(루손섬, 폴리오섬, 카탄도우아네스섬, 마린두께섬, 민도루섬)
Philippines(Luzon I., Polillo I., Catanduanes I., Marindotsuku I., Mindoro I.)

♀

필리핀 민다나오 산 22mm
(Mt. Apo, Mindanao I., Philippines. 2007. 7.)

♂

필리핀 레이테 산 23mm
(Leyte I., Philippines. 2007. 7.)

## 꽃무지아과 / Cetoniinae

### Sternoplus

| 학 명 Scientific name | 채집국 Collected locality | 크 기 Size | 참 고 Remark |
|---|---|---|---|
| 샤우미꽃무지<br>*Sternoplus schaumii* White, 1856 | 인도네시아 Indonesia | ♂♀ 22.5-28.2mm | |

인도네시아 술라웨시 섬에 서식한다. 이 종의 몸빛은 한 가지도 같은 것이 없을 정도로 다양하여 검정색, 녹색, 적갈색 등의 바탕에 노란색의 띠무늬를 가지고 매우 많은 색깔과 무늬의 변이가 어우러지는 무척 재미있는 종이다.

♂

술라웨시 섬 산 26mm
(Camba, S. Sulawesi, Indonesia. 2004. 7.)

♀

술라웨시 섬 산 26mm
(Camba, S. Sulawesi, Indonesia. 2004. 7.)

# Flower Beetles / Tribe 6. Cetoniini

This species, having yellow stripes, inhabits Sulawesi Island, Indonesia. It is interesting as its body color (black, green, red-brown and so on) and pattern show such diversified variations that no individual looks like another.

[Distribution] Sulawesi Island

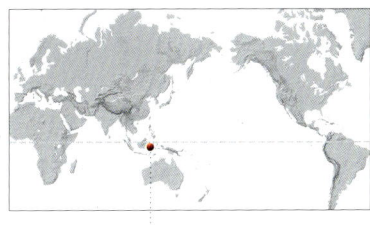

♀

술라웨시 섬 산 26mm
(Bantimurung, S. Sulawesi, Indonesia. 2002. 7.)

♂

술라웨시 섬 산 26mm
(Bantimurung, S. Sulawesi, Indonesia. 2002. 7.)

# 7. 모가슴꽃무지족(族)

이 족은 30속 200종이 알려져 있으며 대부분 북아메리카인 신북구(Nearctic region)에 서식한다. 동양 열대구(Oriental region)에 1속 30종, 에티오피아구(Ethiopian region)에 4속 10여종에 불과하다.
이 장에서는 1속(genus) 2종(species) 3개체를 수록하였다.

This tribe is divided into 200 species under 30 genera, most of them inhabiting the Nearctic region in North America. Currently, 30 species of one same genus live in the Oriental region while some 10 species under 4 genera inhabits the Ethiopian region.
3 individuals of 1 genus, 2 species are included in this section.

Section 7
# Gymnetini

## 꽃무지아과 / Cetoniinae

**Gymnetis**

| 학 명 Scientific name | 채집국 Collected locality | 크 기 Size | 참 고 Remark |
|---|---|---|---|
| 남미점박이모가슴꽃무지<br>*Gymnetis pantherina* White, 1856 | 에콰도르<br>Ecuador | ♂♀ 20-29mm | |

*Gymnetis* 속은 중앙아메리카에서 남아메리카 중부까지 넓게 서식하고 있다. 이 종의 몸빛은 마치 벨벳과 같이 부드러운 황갈색의 미모로 덮여 있으며 광택이 없는 황색바탕에 많은 검정색의 점무늬를 가지고 있다. 방사형 점무늬의 변이가 다양하다.

The *Gymnetis* genus have an expansive habitat from central Africa to the central region of South America. This species flat-yellow bodies, dotted with numerous black speckles, are covered with fine and soft yellowish-brown hair like velvet. The radial-shaped spotted pattern shows a wide range of variations..

[Distribution] Ecuador

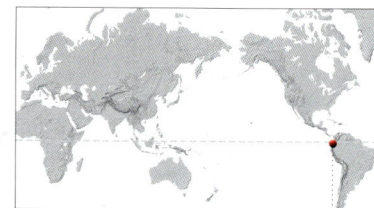
● 채집지 Collected Locality
● 분포지 Distribution

에콰도르 Ecuador

♀

에콰도르 산 20mm
(Ecuador. 2005. 7.)

# Flower Beetles / Tribe 7. Gymnetini

## Gymnetis

| 학 명 Scientific name | 채집국 Collected locality | 크 기 Size | 참 고 Remark |
|---|---|---|---|
| 지카니남미점박이모가슴꽃무지<br>*Gymnetis pantherina zikani* Moser, 1921 | 브라질<br>Brazil | ♂♀ 20-28mm | |

수컷의 복면은 황갈색의 미모로 덮여 있으며 암컷의 복마디는 미모가 없고 검정색을 띠고 있다.

Males` back is covered with some fine yellowish-brown hair while females` abdominal segment, with no hair, is black

[Distribution] Guatemala, Ecuador, Venezuela, Peru, Brazil, Bolivia

♀
브라질 산 20mm
(Rio Sapucai, Minas Gerais, Brazil. 2005. 7.)

♂
브라질 산 21mm
(Rio Sapucai, Minas Gerais, Brazil. 2005. 7.)

## 8. 얼룩꽃무지족(族)

이 족은 21속 100여종이 알려져 있다. 대부분 에티오피아구(Ethiopian region)에 속해 있는 아프리카에 서식하고 있으며 아시아에는 1속 1종만이 알려져 있다.

This tribe is known to be classified into some 100 species under 21 genera. Most of them inhabit the Ethiopian region and other African areas while only 1 species lives in Asia.

## 9. 투구꽃무지족(族)

이 족은 4속 50여종이 알려져 있으며 동양 열대구(Oriental region)에 속해있는 동남아시아에서만 서식한다.
이 장에서는 1속(genus) 5종(species) 11개체를 수록하였다.

This tribe is divided into approximately 50 species under 4 genera, all of which inhabit Southeast Asia, including the Oriental region.
11 individuals of 1 genus, 5 species are included in this section.

  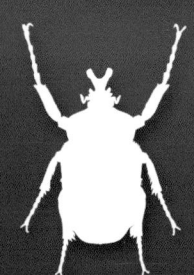 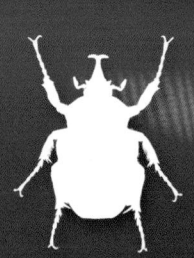 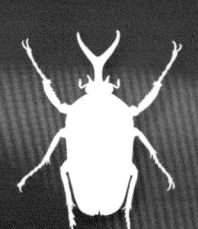 

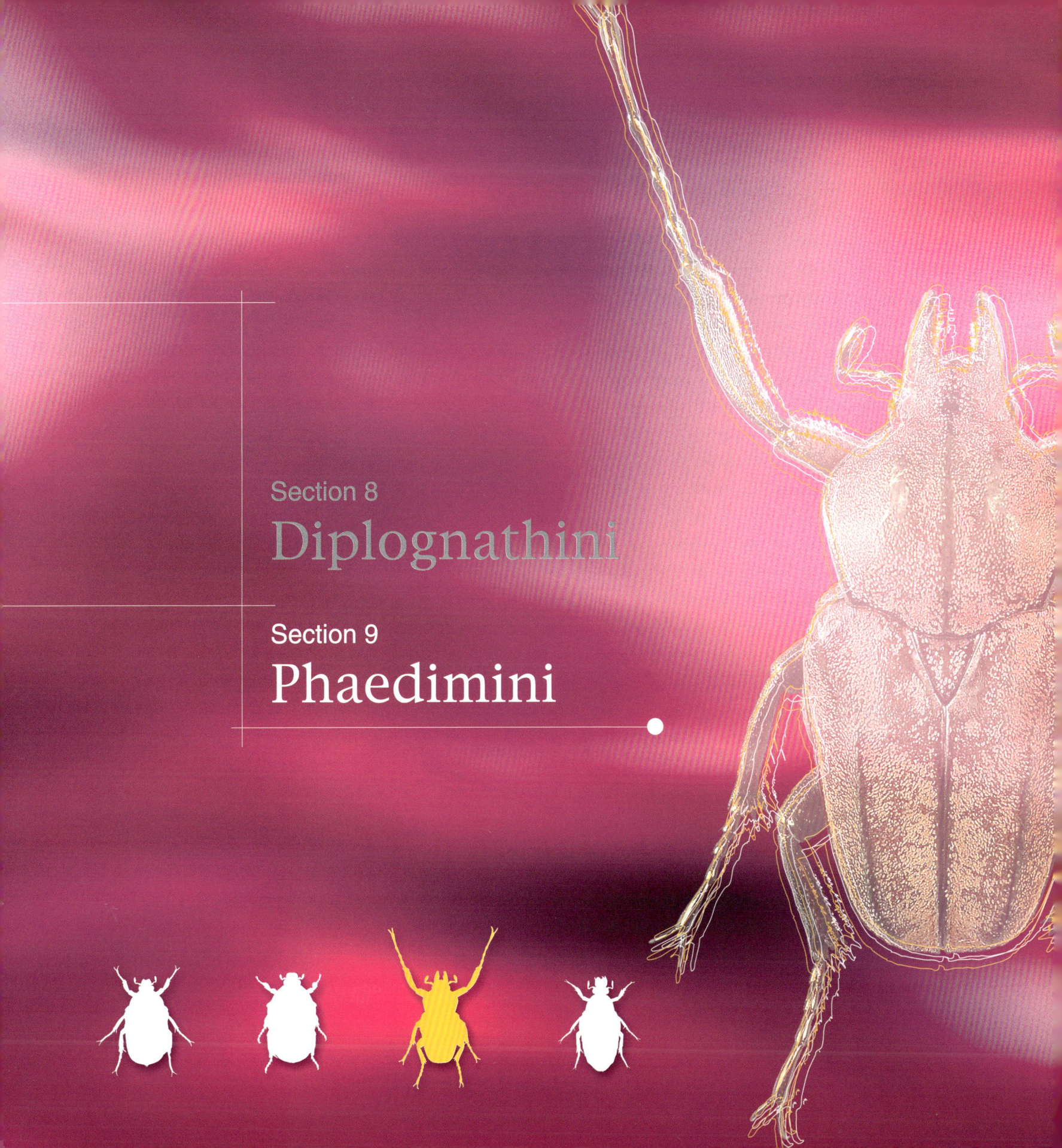

Section 8
# Diplognathini

Section 9
# Phaedimini

# 꽃무지아과 / Cetoniinae

## Mycteristes

| 학 명 Scientific name | 채집국 Collected locality | 크 기 Size | 참 고 Remark |
|---|---|---|---|
| 르히노필루스투구꽃무지<br>*Mycteristes rhinophyllus* Wiedemann, 1823 | 인도네시아<br>Indonesia | ♂ 18-23mm, ♀ 17-19mm | |

인도네시아 자바 섬과 발리 섬에 서식한다. 마치 장수풍뎅이(rhinoceros beetle)와 같이 머리뿔이 발달하여 안쪽으로 굽어 있으며 말단부는 역삼각형이다. 한 개의 가슴뿔은 전방향으로 발달하여 있다. 몸빛은 금동색의 금속성 빛깔을 띠며 미모가 나있다. 앞다리가 길게 발달하여 있다.

They inhabit Java Island and Bali Island, Indonesia. Similar to rhinoceros beetles, they have well-developed and inward-bent cephalic horns, the tips of which are triangular. One cephalic horn is developed in the forward direction. Their golden-copper body is metallic and covered with fine hair. They have well-developed forelegs.

[Distribution] Java Island, Bali Island

동 자바 산 19.2mm
(E. Java, Indonesia. 2004. 4.)

# Flower Beetles / Tribe 9. Phaedimini

## Mycteristes

| 학 명 Scientific name | 채집국 Collected locality | 크 기 Size | 참 고 Remark |
|---|---|---|---|
| 티베타나두뿔투구꽃무지<br>*Mycteristes tibetana* Janson, 1917 | 타이<br>Thailand | ♂ 20-27mm, ♀ 17-21mm | |

티벳, 미얀마, 타이 북부에 서식한다. 몸빛은 금동색이며 가슴은 녹색의 미모로 덮여있다. 머리뿔은 두 개로 발달되어 말단부는 서로가 마주보게 굽어있다. 딱지날개에는 거친 미모가 나있다.

They inhabit Tibet, Myanmar and northern Thailand. Their body color is golden-copper and the thorax is covered with green fine hair. The tips of the two cephalic horns are bent facing each other. The elytra (hard wings) have some rough and fine hair.

[Distribution] Tibet, Myanmar, Thailand

♂  
타이 북부 산 19mm  
(Chiang-Rai, N.Thailand. 2002. 5)

♂  
타이 북부 산 21mm  
(Chiang-Rai, N.Thailand. 2002. 5)

## 꽃무지아과 / Cetoniinae

**Mycteristes**

| 학 명 Scientific name | 채집국 Collected locality | 크 기 Size | 참 고 Remark |
|---|---|---|---|
| 스쿠아모수스뾰족투구꽃무지<br>*Mycteristes squamosus* Ritsema, 1879 | 말레이시아<br>Malaysia | ♂ 20-31mm, ♀ 20-23mm | |

말레이 반도와 수마트라 섬에 서식한다. 몸빛은 녹색의 금속성 광택을 지니고 있으며 황색의 미모로 덮여있다. 수컷의 두 개의 머리뿔은 마치 베트맨의 마스크처럼 뾰족하게 발달하였다.

They inhabit the Malay Peninsula and Sumatra Island. The metallic-green and glossy body is covered with yellow and fine hair. The two cephalic horns of males are pointed like the Batman's mask.

[Distribution] Malay peninsular, Sumatra Island

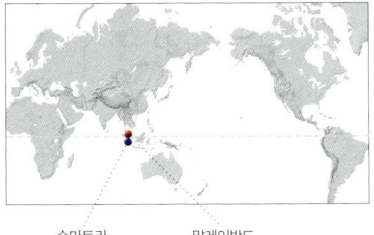

♂

말레이 반도 산 28mm
(Cameron Highlands, W. Malaysia. 2004. 4.)

♀

말레이 반도 산 23mm
(Cameron Highlands, W. Malaysia. 2004. 4.)

Flower Beetles / Tribe 9. Phaedimini

## Mycteristes

| 학 명 Scientific name | 채집국 Collected locality | 크 기 Size | 참 고 Remark |
|---|---|---|---|

볼렌호베니뾰족투구꽃무지
*Mycteristes vollenhoveni* Mohnike, 1871

인도네시아
Indonesia

♂ 28-30mm, ♀ 24-26mm

자바섬에 서식한다. 몸빛은 금녹색의 금속성 광택을 지니고 있으며 황색의 미모로 덮여있다. 수컷의 두 개의 머리뿔은 *M. squamosus*에 비해 좀 더 길다.

Inhabiting Java Island, it has a golden and green, metallic-glossy body covered with fine yellow hair. The males' two cephalic horns are longer than those of *M. squamosus*.

[Distribution] Java Island

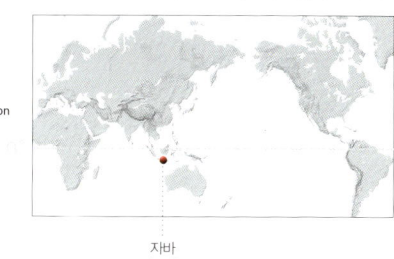

● 채집지 Collected Locality
● 분포지 Distribution

자바
Java I.

♀

동 자바 산 24mm
(Mt. Argopuro, E. Java, Indonesia. 2005. 6.)

♂

동 자바 산 28mm
(Mt. Argopuro, E. Java, Indonesia. 2005. 6.)

## 꽃무지아과 / Cetoniinae

**Phaedimus**

| 학 명 Scientific name | 채집국 Collected locality | 크 기 Size | 참 고 Remark |
|---|---|---|---|
| 하우드니투구꽃무지<br>*Phaedimus howdeni* Arnaud, 1987 | 필리핀<br>Philippines | ♂ 22-26mm, ♀ 19-21mm | |

필리핀 네그로스 섬에 서식한다. 머리와 가슴뿔은 마치 장수풍뎅이처럼 발달되어 있다. 몸빛은 금속성 광택의 녹색으로 딱지날개는 황등색의 넓은 띠를 가지고 있다. 머리뿔은 발달되어 안쪽으로 굽어있으며 말단부는 두 갈래로 나뉘어져 있다. 소형의 개체에서는 말단부가 갈라지지 못하여 부채모양을 하고 있다.

♂

필리핀 네그로스 산 25mm
(Mt. Canlaon, N. Negros I., Philippines 2006. 5.)

♀

필리핀 네그로스 산 21mm
(Mt. Canlaon, N. Negros I., Philippines 2006. 5.)

# Flower Beetles / Tribe 9. Phaedimini

They inhabit Negros Island, Philippines. The cephalic horns and pronotum horns are well-developed like those of rhinoceros beetles. Their body is green and metallic-glossy while the orange-yellow elytra have some wide stripes. he cephalic horns are well-developed and bent inward while the tips are divided into two. Some small individuals` cephalic horn tips, not divided, resemble a fan.

[Distribution] Philippines(Negros Island)

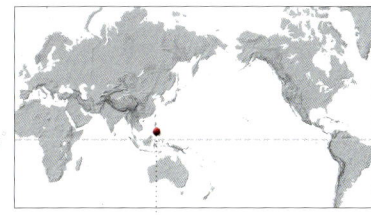

필리핀(네그로스 섬)
Philippines(Negros I.)

필리핀 네그로스 산 22mm
(Mt. Canlaon, N. Negros I., Philippines 2006. 5.)

필리핀 네그로스 산 24mm
(Mt. Canlaon, N. Negros I., Philippines 2006. 5.)

## 10. 홀쭉꽃무지족(族)

이 족은 30속 210종이 알려져 있으며 대부분 동양 열대구(Oriental region)에 속해있는 동남아시아에 서식한다.
이 장에서는 5속(genus) 10종(species) 18개체를 수록하였다.

This tribe is classified into 210 species under 30 genera, most of them inhabiting Southeast Asia, including the Oriental region.
18 individuals of 5 genus, 10 species are included in this section.

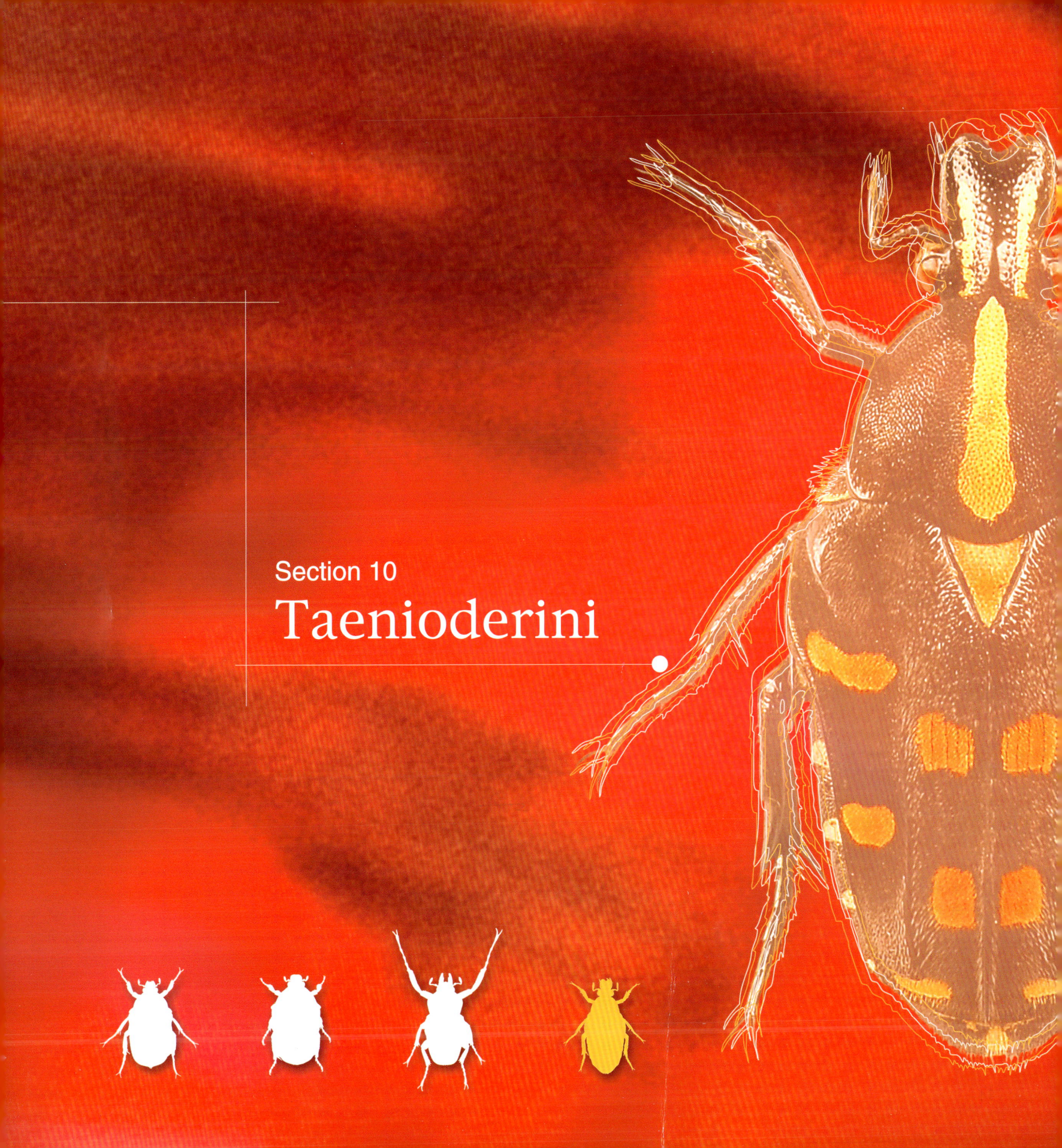

Section 10
# Taenioderini

## 꽃무지아과 / Cetoniinae

| 학 명 Scientific name | 채집국 Collected locality | 크 기 Size | 참 고 Remark |
|---|---|---|---|
| 얼룩홀쭉꽃무지<br>*Euselates* sp. | 필리핀<br>Philippines | ♂ 17-26mm, ♀ 19-21mm | |

필리핀에서 서식한다. *Euselates*속은 43여종이 알려져 있다. 이 속은 딱지날개 양쪽의 가장자리가 안쪽으로 길게 움푹 들어가 홀쭉해진 특징을 가진다. 이 종의 몸빛은 광택이 없는 검정색의 얼룩무늬가 있다. 가슴은 둥그렇고 좁아 바구미를 닮아있다.

They inhabit Philippines. 43 species are known to fall under the *Euselates* genus. This genus is characterized by its elytra (hard wings) whose slim edges reach deep inside. This species has some speckles on its flat-black body. The thorax, round and narrow, resembles a weeril.

[Distribution] Philippines

- 채집지 Collected Locality
- 분포지 Distribution

필리핀 Philippines

♂

필리핀 산 17mm
(Philippines, 2006.)

# Flower Beetles / Tribe 10. Taenioderini

## Euselates

| 학 명 Scientific name | 채집국 Collected locality | 크 기 Size | 참 고 Remark |
|---|---|---|---|
| 스틱티카홀쭉꽃무지<br>*Euselates stictica* Hope, 1847 | 필리핀<br>Philippines | ♂♀ 15-19mm | |

필리핀의 여러 섬에 서식한다. 이 종의 몸빛은 광택이 없는 검정색이며 딱지날개에는 백색의 작은 점들이 있으며 흰색털이 나있다.

This species, which inhabits the Philippines, has a flat black body with some minute spots and white hair on the elytra (hard wings).

[Distribution] Philippines(Luzon, Leyte, Mindanao Island)

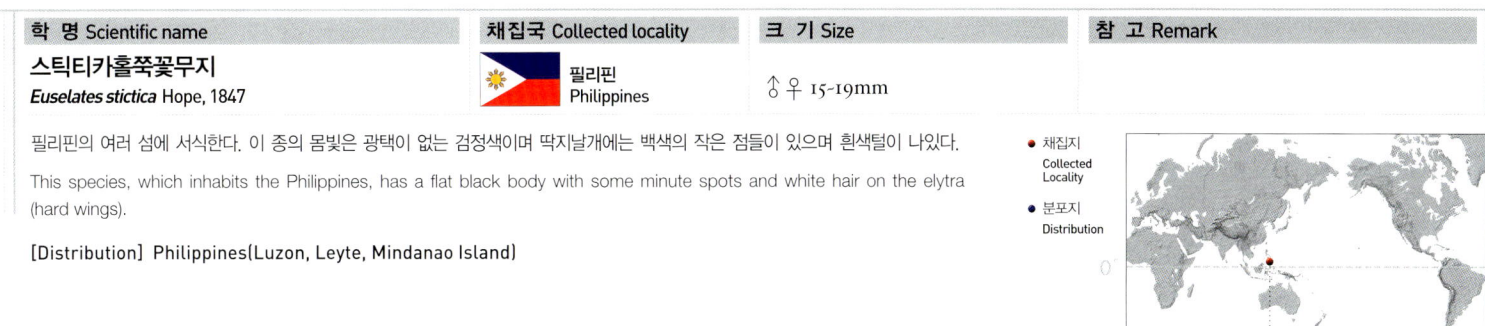

필리핀(루손 섬, 레이테 섬, 민다나오 섬)
Philippines(Luzon I., Leyte I., Mindanao I.)

♀

필리핀 민다나오 섬 산 19mm
(Mt. Syniop, Mindanao I., Philippines, 2006. 6.)

## 꽃무지아과 / Cetoniinae

**Ixorida**

| 학 명 Scientific name | 채집국 Collected locality | 크 기 Size | 참 고 Remark |
|---|---|---|---|
| 엘레강스점박이홀쭉꽃무지<br>*Ixorida elegans* Heller, 1896 | 인도네시아<br>Indonesia | ♂♀ 17-22mm | |

인도네시아 술라웨시 섬에 서식한다. *Ixorida*속은 62여종이 알려져 있다. 이 속 또한, 딱지날개 양쪽의 가장자리가 안쪽으로 길게 움푹 들어가 홀쭉해진 특징을 가진다. 이 종의 몸빛은 광택이 없는 검정색에 황색반점을 가지고 있으며 가슴에 황색의 세로줄을 가지고 있다.

The *Ixorida* genus, which inhabits Sulawesi Island, is divided into 62 species. This genus is characterized by its elytra (hard wings) whose slim edges reach deep inside. This species has some yellow speckles on its flat black body, while there are some yellow stripes on the thorax.

[Distribution] Sulawesi Island

♂

술라웨시 산 22mm
(Palopo, Sulawesi, Indonesia. 2006. 3.)

# Flower Beetles / Tribe 10. Taenioderini

## Ixorida

| 학 명 Scientific name | 채집국 Collected locality | 크 기 Size | 참 고 Remark |
|---|---|---|---|
| 후리데리씨금박무늬홀쭉꽃무지<br>*Ixorida friderici* Heller, 1897 | 인도네시아<br>Indonesia | ♂♀ 17-20mm | |

인도네시아 술라웨시 섬에 서식한다. 이 종의 몸빛은 광택이 없는 검정색에 황색반점을 가지고 있으며 가슴에 황색의 세로줄을 가지고 있다.

This species, which inhabits Sulawesi Island. This species has some yellow speckles on its flat black body, while there are some yellow stripes on the thorax.

[Distribution] Sulawesi Island

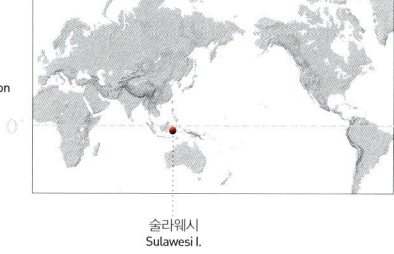

술라웨시
Sulawesi I.

♂

술라웨시 산 18.5mm
(Mt. Pedamaran, Sulawesi, Indonesia. 2005. 7.)

♂

술라웨시 산 19mm
(Mt. Pedamaran, Sulawesi, Indonesia. 2005. 7.)

# 꽃무지아과 / Cetoniinae

## Ixorida

| 학 명 Scientific name | 채집국 Collected locality | 크 기 Size | 참 고 Remark |
|---|---|---|---|
| 아펠레스주홍줄홀쭉꽃무지<br>*Ixorida venerea apelles* J. Thomson, 1860 | 인도네시아<br>Indonesia | ♂♀ 15-17mm | |

인도네시아 말루쿠 제도에 서식한다. *Ixorida*속은 62여 종이 알려져 있다. 몸빛은 가슴이 검정색이고 나머지는 적갈색이며 약한 광택을 가지고 있다. 머리에 두 개의 세로줄, 가슴에 세 개의 세로줄, 딱지날개에 서로 연결되지 않은 3쌍의 세로줄 무늬는 짙은 황색의 미모로 덮여있다.

The *Ixorida* genus, which inhabits Maluku Island, is divided into 62 species. The thorax is black while the rest of the body is reddish-brown, subtly glossy. The two cephalic stripes, three on the thorax and another six, disconnected on the elytra (hard wings), are covered with fine dark-yellow hair.

[Distribution] Indonesia(Maluku Islands)

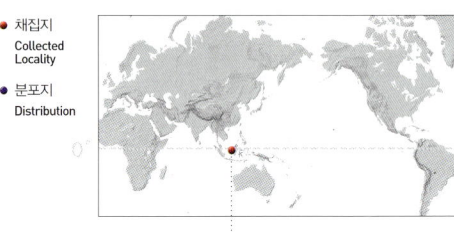

인도네시아(말루쿠 제도)
Indonesia (Maluku Is.)

♂              ♀

인도네시아 말루쿠 제도 산 16mm      인도네시아 말루쿠 제도 산 15.5mm
(Maluku Is., Indonesia. 2006. 7.)      (Maluku Is., Indonesia. 2006. 7.)

# Flower Beetles / Tribe 10. Taenioderini

## Ixorida

| 학 명 Scientific name | 채집국 Collected locality | 크 기 Size | 참 고 Remark |
|---|---|---|---|
| 솔로모니카주홍줄홀쭉꽃무지<br>*Ixorida solomonica* Miksic, 1972 | 솔로몬<br>Solomon | ♂♀ 14-17mm | |

솔로몬 제도에 서식한다. *I.v.apelles.*와 비교하여 몸빛은 전체가 검정색이고 나머지는 적갈색이며 보다 약한 광택을 가지고 있다. 머리에 두 개의 세로줄, 가슴에 세 개의 세로줄, 딱지날개에 서로 연결되지 않은 3쌍의 세로줄 무늬는 옅은 황색의 미모로 덮여있다.

They inhabit Solomon Islands. Compared with that of *I.v.apelles*, their body, black and sometimes reddish-brown, has a subtler gloss. The two cephalic stripes, three on the thorax and another six, disconnected on the elytra (hard wings), are covered with fine light-yellow hair.

[Distribution] Solomon Islands (Bougainville Island, Malaita Island, Guadalcanal Island, Englagi Island)

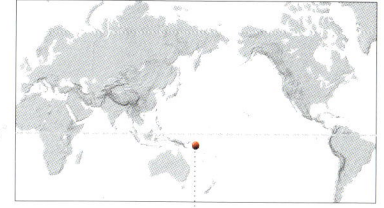

솔로몬 제도(부건이루 섬, 말라이타 섬, 가다루카나루 섬, 엥라기 섬)
Solomon Is. (Bougainville I., Malaita I., Guadalcanal I., Englagi I.)

♀

솔로몬 제도 산 14mm
(Malaita I., Solomon Is.. 2006. 7.)

♂

솔로몬 제도 산 14mm
(Malaita I., Solomon Is.. 2006. 7.)

# 꽃무지아과 / Cetoniinae

## Ixorida

| 학 명 Scientific name | 채집국 Collected locality | 크 기 Size | 참 고 Remark |
|---|---|---|---|
| 프로핀쿠아흰점홀쭉꽃무지<br>*Ixorida propinqua* Mohnike, 1873 | 필리핀<br>Philippines | ♂♀ 16-20mm | |

필리핀의 민다나오 섬에 서식한다. 이 종은 검정색 바탕에 광택을 지니고 있으며 가슴과 딱지날개에 흰 점무늬를 가지고 있다. 수컷은 거친 미모가 있으나 암컷은 없다.

This species, which inhabits Mindanao Island, Philippines, has a glossy and flat black body with some white speckles on the thorax and elytra (hard wings). Males have rough, fine hair unlike their female counterparts.

[Distribution] Philippines(Mindanao Island)

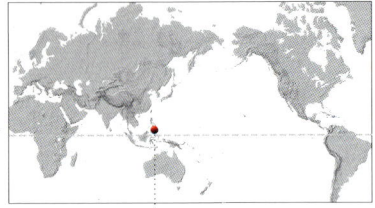

필리핀(민다나오 섬)
Philippines(Mindanao I.)

♂

필리핀 민다나오 산 19.2mm
(Mindanao I., Philippines. 2006. 7.)

♀

필리핀 민다나오 산 18.2mm
(Mindanao I., Philippines. 2006. 7.)

Flower Beetles / Tribe 10. Taenioderini

## Plectrone

| 학 명 Scientific name | 채집국 Collected locality | 크 기 Size | 참 고 Remark |
|---|---|---|---|
| 앤드로애디홀쭉꽃무지<br>*Plectrone endroedii* Miksic, 1974 | 필리핀<br>Philippines | ♂♀ 24.0-29.3mm | |

필리핀의 민다나오 섬에 서식한다. 이 종은 고광택의 금속성 색감을 지니고 있다. 녹색, 청색, 검정색 등의 색깔을 가지고 있으며 금속 세공품을 보는 듯하다.

This species, which inhabits Mindanao Island, Philippines, is highly-glossy and metallic. Its various colors (green, blue, black and so on) are like a cafted matal piece.

[Distribution] Philippines (Mindanao, Samar, Leyte Island)

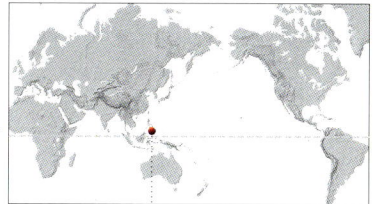

필리핀(민다나오 섬, 사마이루 섬, 레이트 섬)
Philippines (Mindanao I., Samar I., Leyte I.)

♀

필리핀 민다나오 섬 산 28mm
(Mindanao I., Philippines. 2007. 5.)

♂

필리핀 민다나오 섬 산 29mm
(Mindanao I., Philippines. 2007. 5.)

# 꽃무지아과 / Cetoniinae

## Plectrone

| 학 명 Scientific name | 채집국 Collected locality | 크 기 Size | 참 고 Remark |
|---|---|---|---|
| 루마위기검은박쥐꽃무지<br>*Plectrone lumawigi* Miksic, 1986 |  필리핀<br>Philippines | ♂♀ 26-31mm | |

필리핀의 파나이(Panay) 섬에 서식한다. 이 종은 검정색 바탕에 금속성 광택을 지니고 있으며 머리와 가슴, 딱지날개 양쪽은 마치 금속을 세공하여 놓은 것과 같이 점각이 밀집되어 있다.

This species, which inhabits Panay Island, Philippines, has a glossy and flat body black in color. The cephalic part, thorax and both sides of the elytra (hard wings) have some concentrated stipples like a crafted metal piece.

[Distribution] Philippines(Panay Island)

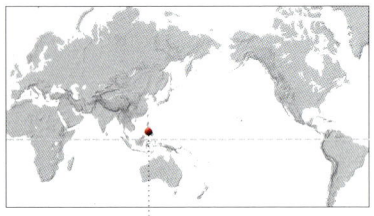

● 채집지 Collected Locality
● 분포지 Distribution

필리핀(파나이 섬)
Philippines(Panay I.)

♂
필리핀 파나이 산 28mm
(Panay I., Philippines 2007. 10.)

♀
필리핀 파나이 산 29mm
(Panay I., Philippines 2007. 10.)

# Flower Beetles/ Tribe 10. Taenioderini

## Taeniodera

| 학 명 Scientific name | 채집국 Collected locality | 크 기 Size | 참 고 Remark |
|---|---|---|---|
| 트리칼라홀쭉꽃무지<br>*Taeniodera tricolor tricolor* (Mohnike, 1873) | 필리핀<br>Philippines | ♂ ♀ 18-25mm | |

필리핀 전역에 걸쳐 서식한다. 이 종은 검정색 바탕을 포함하여 주홍색과 노란반점의 세가지 색으로 이루어져 있다. 머리와 다리는 미약한 광택을 가지고 있지만 가슴과 딱지날개는 광택이 없다.

This species lives across the Philippines. Its black body is dotted with some scarlet or yellow speckles. The cephalic part and legs have a subtle gloss while the thorax and elytra (hard wings) have none.

[Distribution] Philippines (Luzon Island, Mindoro Island, Panay Island, Negros Island, Mindanao Island)

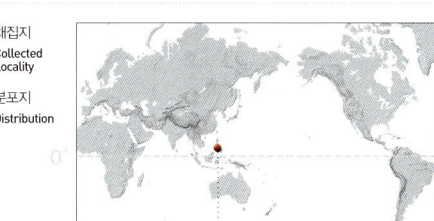

● 채집지 Collected Locality
● 분포지 Distribution

필리핀(루손 섬, 민도르 섬, 파나이 섬차나이 섬, 네그로스 섬, 민다나오 섬)
Philippines (Luzon I., Mindoro I., Panay I., Negros I., Mindanao I.)

♀

필리핀 네그로스 산 20mm
(Mt. Kiandana, Negros I., Philippines 2008. 3.)

♂

필리핀 네그로스 산 23mm
(Mt. Kiandana, Negros I., Philippines 2008. 3.)

# 도감의특징
## Special Point of this Book

곤충의 세부적인 부분과 질감의 느낌을 그동안 표현할 수 없었던 어려움을 이 도감에서는 일반적인 눈으로 관찰 할 수 없는 부분도 고해상도로 1.1배에서 최대 12배까지 배율을 확대하여 촬영함으로써 마치 광학정밀 입체현미경을 통하여 보는 것과 같은 효과를 구현하였다.

Most of books published until now were very lacking in expressing the detailed shapes and textures of insects. In order to address these issues in this book, sections that are difficult to observe with the naked eye were magnified from 1.1 to 12 times using high resolution proximity photography, in order to create an effect as if the reader was viewing it through a stereo microscope.

# 사진제작과정
## Photo Produce Process

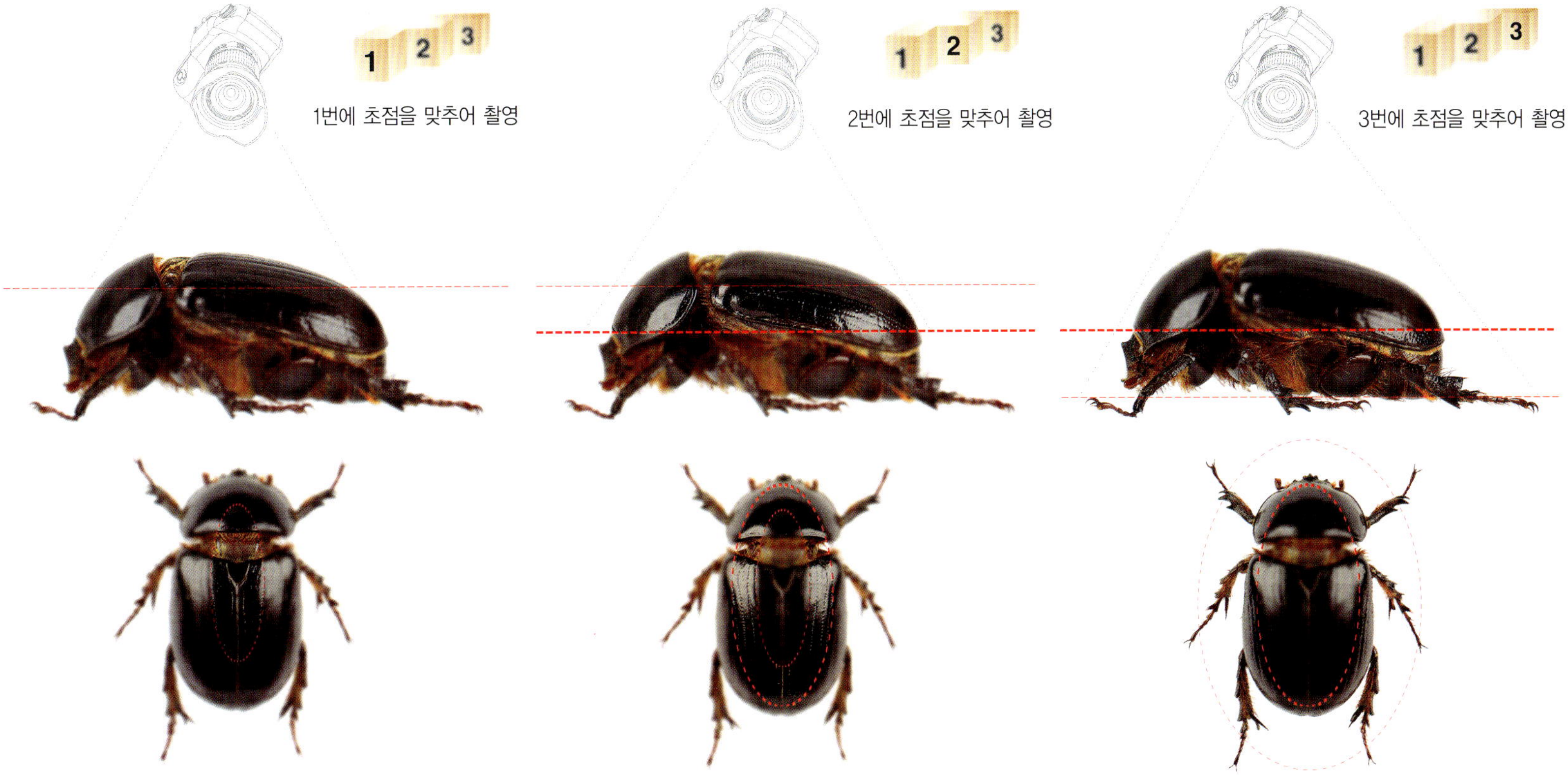

1번에 초점을 맞추어 촬영     2번에 초점을 맞추어 촬영     3번에 초점을 맞추어 촬영

## 부분접사 및 심도합성

카메라의 렌즈는 사람의 눈과 같이 가까운 거리에서 물체를 크게 보려고 하면 자세히는 보이나 초점의 깊이가 얇다. 이것을 카메라에서는 심도라고 하는데 확대하여 접사 할수록 초점이 맞는 부위가 적어진다. 따라서 전체의 초점을 맞추기 위하여 부분적으로 서로 다르게 촬영하여 사진을 겹침으로 전체의 초점이 맞는 사진을 구현할 수 있게 했다.

이와 같은 원리를 이용하여 이 도감에 실려있는 사진들은 3장에서 최대 9장까지 서로 다른 초점부위를 촬영하여 컴퓨터그래픽을 통하여 합성하고, 색상보정과 덧칠로 곤충의 세부적인 부분과 질감을 표현하였다. 이처럼 본 도감의 가장 큰 특징은 바로 고해상도 사진제작에 있다.

※ 위의 그림은 실제 작업과정과 차이가 있지만 이해를 돕기 위해 표현하였다.

## Partial Close-Up and Pictorial Synthesis

The lens of a camera is like human eyes as its focus gets shallower as the object being watched gets nearer. This is called the depth of a camera, meaning that the closer a camera gets to its object, the blurrier the focus gets. This book, taking into consideration such property, overlapped some partially-shot pictures to adjust the entire focus.

To be more specific, for each picture in this book, three to nine pictures of different focuses were synthesized by computer graphics, color-calibrated and recoated to express the details and texture of beetles.

One of the biggest advantages of this book is the high-resolution pictures made this way.

※The picture above, though somewhat different from the actual procedure, is to help increase the readers' understanding.

일반접사촬영사진
Ordinary Close-Up

부분접사촬영심도합성사진
Partial Close-Up and Synthesis

# 저자소개

서울 천호초등학교 졸업
서울 천호중학교 졸업
서울 성동고등학교 졸업
우석대학교 농학부 낙농학과 졸업
우석대학교 대학원 생물학과 이학석사
우석대학교 대학원 생명공학과 박사과정 입학
원광대학교 대학원 동물응용과학 박사과정
우석대학교 야생생물연구회 고문
원광대학교 동물자원개발연구센터 고문
한국매개동물치료협회 이사
한국애완반려동물학회 이사
원광대학교 생명자원과학대 애완동식물학부 강사
부안군 지역개발단 부안곤충탐사과학관 건립담당자

손민우 (孫敏又) Son Minwoo  1965.7.22

# About the Author

| | |
|---|---|
| 1986년 9월 | 전북의 야생화와 나비전시회 |
| 1989년 9월 | 한국의 야생생물전시회 |
| 1992년 9월 | 한국 곤충전시회 |
| 1993년 9월 | 한국 곤충생태전시회 |
| 1995년 9월 | 한국의 곤충 및 해양표본전시회 |
| 2004년 10월 | 전주 세계소리축제 곤충소리 특별전(한국소리문화전당) |
| 2004년 12월 | 로봇곤충전 곤충전시회(광주비엔날레) |
| 2005년 7월 | EBS와 함께하는 세계 곤충학습체험전(세종문화회관 본관) |
| 2006년 3월 | 학습용곤충표본에 대한 실용신안 등록(2005-0034633) |
| 2006년 12월 | 학습용곤충표본에 대한 발명특허 등록(10-0644324) |

## 저서 Literary Works

**동물매개치료**   손민우 외 8명 공저 - 학지사 2007년 7월 10일

**동물영양관리사양학**   손민우/김옥진 공저 -도서출판 선진 2008년 3월 3일

**세계유용곤충대도감시리즈 Vol. 1 세계의 사슴벌레 대도감**   손민우 저 - (주)커뮤니케이션열림 2008년 4월 10일

**동물번식생리학**   손민우/김옥진 공저 - 선진출판기획 2008년 9월 1일

**세계유용곤충대도감시리즈 Vol. 2 세계의 장수풍뎅이 대도감**   손민우 저 - (주)커뮤니케이션열림 2009년 4월 18일

**세계유용곤충대도감시리즈 Vol. 3 세계의 꽃무지 대도감**   손민우 저 - (주)커뮤니케이션열림 2009년 4월 18일

**세계유용곤충대도감시리즈 Vol. 4 세계의 하늘소 대도감**   손민우 - 집필중

## 논문 Thesis

**애완곤충을 이용한 매개치료의 기대효과**   Animal Assisted Therapy using Pet Insects

손민우 외 2명 - Receive 29 March 2007 / Accepted 31 May 2007

**가거도 해역의 태형동물의 분류학적 연구**   Taxonomic Study on Bryozoans from Gageodo Is. Waters

손민우 - Receive 19 October 2007 / Accepted 22 November 2007

**애완곤충을 이용한 매개·치료의 기대효과II (애완곤충을 이용한 색채테라피의 기대)**

Animal Assisted Therapy Using Pet Insects - The Effect of Color Therapy Using Pet Insects-

손민우/김옥진-Received April 11, 2008 / Accepted May 20, 2008

# 에피소드

# Episode

20여 년에 걸쳐 세계를 누비며 곤충을 채집하고 연구해 온 그는 근래에 곤충과학관을 건립중이며, 그동안 채집한 곤충을 정리하여 독자적으로 개발한 사진촬영 기법과 합성기술을 이용하여 세계최초의 초고해상도 곤충표본도감을 만들고 있다. 그는 지금도 세계 각지를 돌며 후세에게 물려줄 세계의 곤충들을 채집 중이며 곤충의 색상을 이용한 컬러테라피 등 다양한 곤충문화 컨텐츠 개발에 혼신을 다하고 있다.

For about 20 years, going around the world, He has been collect and research the insects. In these days, He construct an Insect Museum and arrange the insects which has collected, And making a Photograph Books insect of the world with hi-density digital photos those are resulted by his own developed photograph process and photo composite methods.

# Episode 1

남중국해 상공   Flight over South China Sea. Feb. 2005.

## 에피소드

타이, 치앙마이
Chiang Mai, N.Thailand. Dec. 16. 2007.

타이 북부, 아카민족의 장승
Totem pole, Akka Village, N. Thailand. Feb. 2001.

타이의 곤충요리
Insects food of Thailand. Feb. 2001.

 동남아 탐사 여행 With My Colleague

풍물 Scenery

동남아의 음식 Foods of Jungle

## 탐사 도중 만난 소수민족 Peoples of Jungle

# Episode

치앙라이 아카민족의 여인
Akka People of Chiang Rai, N. Thailand. Jun. 6. 2004.

## 탐사 도중 만난 아이들 Chileren of Jungle

치앙라이 카렌족의 어린이
Karen People of Chiang Rai, N. Thailand. Jun. 15. 2004.

## 탐사 도중 머물터 Houses of Jungle

수마트라 바탁민족의 가옥
Batak Houses, Lake Toba of Sumatra, Indonesia. Sep. 17. 2003.

171

# 에피소드

치앙마이
Chiang Mai, N. Thailand. Jun. 14. 2004.

방콕의 외곽
Bangkok, Thailand. Feb. 2. 2002.

콰이어강
Kwai River, Kanchanaburi, CW. Thailand. May. 24. 2004.

동남아 곤충 탐사  Looking for Insects

정글  Jungle

타이 중서부
Thailand middle west

# Episode

말레이시아 카메룬 하이랜드 Malaysia

보르네오 Borneo

수마트라 Sumatra

말레이시아 카메룬 하이랜드
Cameron Highland, W. Malaysia. May. 19. 2006.

동남아 최고봉 해발 4093m, 보르네오 코타키나바루
Mt. Kota Kinabalu, Borneo, E. Malaysia. May. 21. 2006.

수마트라 레이크토바, 바탁민족의 장승
Batak, Lake Toba, Sumatra. Indonesia. Sep. 18. 2003.

# 에피소드

치앙라이 라오족의 뱀브보트
Bamboo-boat of Lao People. Chiang Rai, N.Thailand. Dec. 14. 2007.

골든 트라이앵글  Golden Triangle, N. Thailand

중국 쓰촨성 칭청산  해발 1,600m (세계문화유산)
Mt. Qing cheng shan, Sichun Shng, China. July. 25. 2007.

중국 칭청산  Mt. Qing cheng shan, China

타이, 치앙마이
Chiang Mai, N.Thailand. Dec. 16. 2007.

태국 북부  N. Thailand

174

# Episode

티벳 동부  E. Tibet Plat

동티벳 홍위앤 고원 해발 3,500m
E. Tibet Plat. Alt. 3,500m. July. 24. 2007.

전북 부안 변산반도국립공원
Mt. Byeonsan, Korea National Park

전북 부안 변산반도국립공원
Mt. Byeonsan, Korea National Park. May. 27. 2007.

전북 부안 내변산
Mt. Byeonsan, Korea National Park

전북 부안 변산반도국립공원
Mt. Byeonsan, Korea National Park. May. 3. 2007.

## 에피소드

전북 부안 변산반도국립공원
Mt. Byeonsan, Korea National Park. June. 5. 2007.

세계곤충학습체험전, 세종문화회관
The exhibition of learning insects of the world, Sejong Center. July. 15~31. 2005.

집무실에서
Son's Office. September. 1. 2007.

필드 Field

전시회 Exhibition

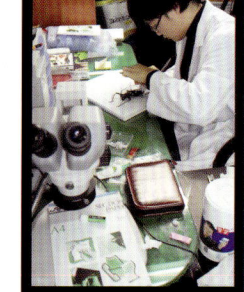

표본작업 및 촬영
Sample & photographing

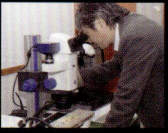
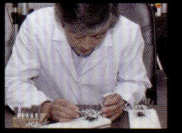
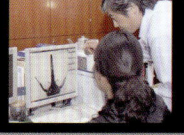

# Episode

편집디자인 Editorial Design

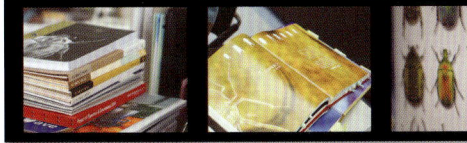

커뮤니케이션 열림 대표 집무실.
Office of Communication Yeollim. April. 01. 2009.

인쇄
Printing

인쇄작업 현장.
Printing Process. April. 06. 2009.

발행인 _ 박철영
Publisher _
Park Cheolyoung

아트디렉션 _ 은성태
Art Direction _
Eun Sungtae

사진합성/트리밍 _ 이수진
Trimming/Montage _
Lee Sujin

편집디자인 _ 임성수
Editorial Direction _
Lim Sungsu

편집디자인 _ 김영준
Editorial Design _
Kim Youngjune

제작관리 _ 이성철
Process Management _
Lee Sungcheol

177

# 도움주신분들
## Thanks to

### 감 수

우석대학교 생물학과 교수　박사 서지은
전 나비학회장 / 성보화학이사　박사 오성환
전 한국조폐공사 디자인실장 / 중부대학교 인쇄미디어학과 교수 오순환

### 도움 주신 분

만천곤충박물관 / 전 국제곤충연구소 회장 김태완
전 한국곤충동호인회 회장 홍승표
일본원서 번역 야마다 요꼬

### Supervisor

Professor of Biology at Woosuk University　Dr. Seo Jieun
Director of Sung-bo Chemicals Co., Ltd. / Former President of Korea Butterfly Research Society　Dr. Oh Sunghwan
Former Art Director of Korea Minting & Seurity Printing Corporation / Professor of Toongbu University Dept. & Print Media　Oh Sunhwan

### Thanks to

Director of Mancheon Insect Museum & International entomological Institute　Kim Taewoan
Former Cheirman of Korea Insect mania Association.　Hong Seungpyo
Japanese Translation　Yamada Yoko

# 참고문헌
## References

外國産クワガタ・カブトムシ飼育大図鑑 / 世界文化社
Endless Collection Series Vol.1 Flower Beetles コレクショソシリーズ・ハナムグリ / 酒井 香 / ESI
Endless Collection Series Vol.3 Giant Beetles Euchirinae・Dynastinae テナガコガネ・カブトムシ / 水沼 哲郎 / ESI
カブトムシの百科 第4版 / 悌鶴野義嗣
原色昆蟲大圖鑑2 / 同社編 / 北隆館
The Cetoniinae Beetles of the World / Mushi-Sha's Iconographic Series of Insects 3
日本産幼虫大図鑑 編集部(編さん) / 学習研究社
原色日本甲虫図鑑(1) / 森本 桂 / 保育社
原色日本甲虫図鑑(2) / 保育社
原色日本甲虫図鑑(3) / 黒沢 良彦 / 保育社
原色日本甲虫図鑑(4) / 保育社
原色昆虫図鑑〈1〉 / 単行本(ソフトカバー) / 北隆館
原色昆虫図鑑〈2〉 / 甲虫他 / 単行本(ソフトカバー) / 北隆館
完璧版 昆虫の写真図鑑―オールカラー 世界の昆虫、クモ、その他の虫300科 地球自然ハンドブック / 日本ヴォーグ社
世界のクワガタ大圖鑑 - エイムック 趣味の昆蟲編集部 / えい出版社
カブト・クワガタ・ハナムグリ300種図 坪井 源幸 / ピーシーズ
昆虫 ニューワイド学研の図鑑 / 学習研究社
世界の昆虫 ニューワイド学研の図鑑 / 学習研究社
世界のクワガタムシ / アリス館
甲虫 山渓フィールドブックス / 山と渓谷社
原色図鑑 野外の毒虫と不快な虫 原色 図鑑 / 全国農村教育協会
世界珍蝶図鑑 熱帯雨林編 / 人類文化社
人氣 外國産 甲蟲 / 海野和男
中國珍稀圖鑑 / 國林業版社
세계유용곤충대도감시리즈 I. 세계의 사슴벌레 대도감 / (주)커뮤니케이션열림
에로이카 자연과학탐구(장수풍뎅이편) / 뉴턴코리아
곤충 용어집 / 한국응용곤충학회, 한국곤충학회
한국의 곤충 / 남상호 / 교학사
한국의 딱정벌레 / 김정환 / 교학사
곤충 / (주)은하수미디어
한국곤충명집 / 건국대학교출판부
강원의 자연 / 강원교육청
한국경제곤충10 풍뎅이 上科 / 농업과학기술원
동물분류학 / 한국동물분류학회
자원곤충학 / 박규태 / 아카데미서적
일반곤충학 / 한국곤충학회 / 법문사
사회과부도 / (주)지학사
제 2판 생물학용어집 / 한국생활과학협회 편
학습원색대도감 곤충 1 / 금성출판사
학습원색대도감 곤충 2 / 금성출판사

# 부안누에타운 곤충과학관 Insect Science Hall of the Buan Silkworm Town

개관 예정일 2009. 10.

## Buan Silkworm Town Insect Science Museum and Insect Ecology Exploration Science Park

'부안곤충탐사과학공원 조성사업'은 2006년 지방과학관 확충사업에 선정되어, 유용곤충 산업의 일환인 부안누에타운 총 396,696㎡ 내, 297,522㎡의 순환형 생태체험장으로 조성됩니다.

부안곤충탐사과학공원은 변산반도국립공원과 함께 자리하여, 이름에서 엿볼 수 있듯이 곤충의 생태를 과학적으로 탐사함과 동시에 체험과 놀이를 즐길 수 있도록 조성됩니다. 건축면적 991.74㎡ 규모의 곤충과학관이 건립되고, 건축면적 330.56㎡에는 유용곤충인 누에의 생태와 산업적 이용을 보여주는 체험관으로 구성됩니다.

곤충과학관은 테마를 보여주는 기획전시실, 사계절 언제든지 즐길 수 있는 체험관, 곤충 표본을 보관하는 수장고, 그리고 과학적 연구를 담당하는 연구실로 이루어집니다.

특히 계절별로 테마를 구성하는 기획전시는 고정된 전시물이 아닌 변화로운 것으로서 부안곤충탐사과학관만이 갖는 전시가 될 것입니다. 그래서 '보고 또 보고, 찾고 다시 찾는' 곤충과학 시설로서 청소년과 어린이들의 기초 과학 교육장으로 성황을 이룰 것으로 기대하고 있습니다.

곤충만이 갖는 색채 및 모양의 다양성, 생활의 다양한 적응력, 산업으로의 무한한 이용성 등은 다양한 감성을 조기에 심어주는 데에 무엇보다 좋은 구성요소가 될 것입니다.

부안곤충탐사과학관의 주요 전시물은 세계 곳곳에서 직접 채집하여 표본으로 삼은 것으로, '세계의 사슴벌레', '세계의 장수풍뎅이', '세계의 하늘소', '세계의 딱정벌레' 등 테마별로 일정한 시간적 간격을 두고 기획 전시 될 것입니다. 특히 학습용 곤충표본은 저자가 20년간 세계를 탐사하며 채집한 곤충으로 이미 2005년도 세종문화회관 본관에서 성공적인 전시를 가졌던 인정받은 작품입니다. '나도 곤충 박사'라는 전시공간은 과학자들이 곤충을 촬영하여 도감을 제작하고, 연구하는 데 사용하는 과학 기자재와 그 방법을 전시하여, 직접 체험할 수 있도록 구성함으로써 인기를 얻을 것으로 기대하고 있습니다.

'사계절 체험 학습장'은 밀림을 축소하여 만든 소품들과 살아 움직이는 곤충으로 구성한 것으로서, 탐사를 위한 과학 기자재를 들고 직접 곤충의 생활사를 지켜볼 수 있는 공간입니다. 유년 및 청소년들에게는 오감을 만족시키는 참여형 학습장으로서 깊은 감동을 주게 될 것으로 기대합니다.

또한 297,522㎡의 규모로 조성될 탐사형 과학 공원은 곤충의 계절별 생태와 우화 과정 등의 변화를 직접 체험할 수 있도록 구성됩니다. 생태지형의 지도와 탐구활동을 위한 메뉴얼을 제공함으로써 각기 다양한 생물들이 '어떻게 적응하여 살고 있는지', '어떻게 서로와 연관을 가지는지' 등을 자연스럽게 체험할 수 있도록 배려하였습니다. 즉, 스스로 답을 찾아 갈 수 있도록 탐사코스를 과학적으로 조성함으로써 관람에만 그치는 것이 아니라, 자연과 하나가 되어 사물과 현상을 과학적으로 풀어가는 현장 학습장이 될 것입니다.

부안누에타운 곤충과학관 조감도

# Insect Exhibition Facility 곤충전시시설

우리나라에는 표본이나 살아 있는 곤충 등, 곤충 모습과 생태를 전시하고 있는 곳들이 있다. 이런 곳에 가면 희귀하거나 신기한 곤충 등을 볼 수 있다. 지도에 각각의 장소를 표시하여 소개한다.

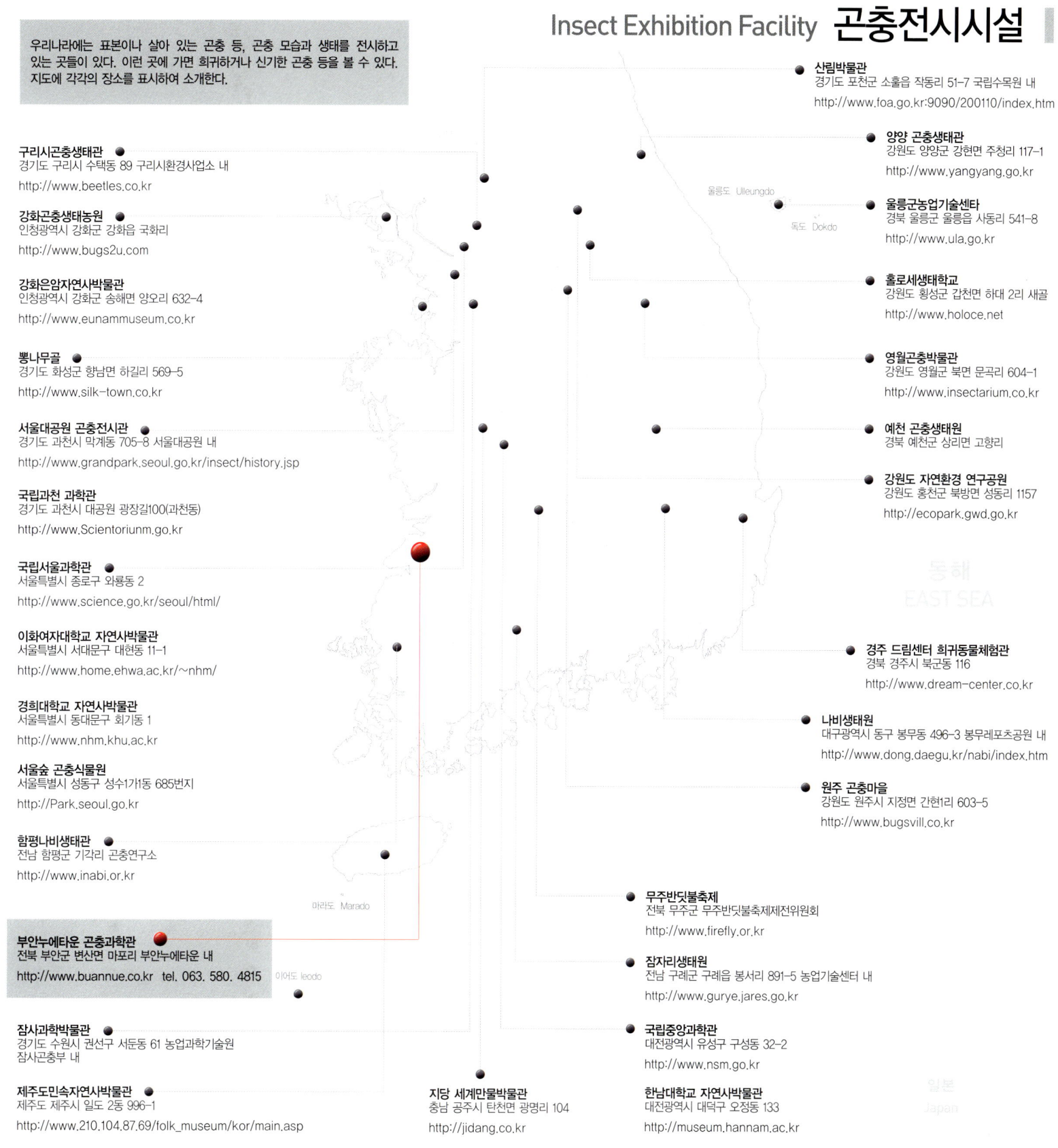

**구리시곤충생태관**
경기도 구리시 수택동 89 구리시환경사업소 내
http://www.beetles.co.kr

**강화곤충생태농원**
인천광역시 강화군 강화읍 국화리
http://www.bugs2u.com

**강화은암자연사박물관**
인천광역시 강화군 송해면 양오리 632-4
http://www.eunammuseum.co.kr

**뽕나무골**
경기도 화성군 향남면 하길리 569-5
http://www.silk-town.co.kr

**서울대공원 곤충전시관**
경기도 과천시 막계동 705-8 서울대공원 내
http://www.grandpark.seoul.go.kr/insect/history.jsp

**국립과천 과학관**
경기도 과천시 대공원 광장길100(과천동)
http://www.Scientoriunm.go.kr

**국립서울과학관**
서울특별시 종로구 와룡동 2
http://www.science.go.kr/seoul/html/

**이화여자대학교 자연사박물관**
서울특별시 서대문구 대현동 11-1
http://www.home.ehwa.ac.kr/~nhm/

**경희대학교 자연사박물관**
서울특별시 동대문구 회기동 1
http://www.nhm.khu.ac.kr

**서울숲 곤충식물원**
서울특별시 성동구 성수1가1동 685번지
http://Park.seoul.go.kr

**함평나비생태관**
전남 함평군 기각리 곤충연구소
http://www.inabi.or.kr

**부안누에타운 곤충과학관**
전북 부안군 변산면 마포리 부안누에타운 내
http://www.buannue.co.kr   tel. 063. 580. 4815

**잠사과학박물관**
경기도 수원시 권선구 서둔동 61 농업과학기술원
잠사곤충부 내

**제주도민속자연사박물관**
제주도 제주시 일도 2동 996-1
http://www.210.104.87.69/folk_museum/kor/main.asp

**산림박물관**
경기도 포천군 소흘읍 작동리 51-7 국립수목원 내
http://www.foa.go.kr:9090/200110/index.htm

**양양 곤충생태관**
강원도 양양군 강현면 주청리 117-1
http://www.yangyang.go.kr

**울릉군농업기술센타**
경북 울릉군 울릉읍 사동리 541-8
http://www.ula.go.kr

**홀로세생태학교**
강원도 횡성군 갑천면 하대 2리 새골
http://www.holoce.net

**영월곤충박물관**
강원도 영월군 북면 문곡리 604-1
http://www.insectarium.co.kr

**예천 곤충생태원**
경북 예천군 상리면 고향리

**강원도 자연환경 연구공원**
강원도 홍천군 북방면 성동리 1157
http://ecopark.gwd.go.kr

**경주 드림센터 희귀동물체험관**
경북 경주시 북군동 116
http://www.dream-center.co.kr

**나비생태원**
대구광역시 동구 봉무동 496-3 봉무레츠공원 내
http://www.dong.daegu.kr/nabi/index.htm

**원주 곤충마을**
강원도 원주시 지정면 간현1리 603-5
http://www.bugsvill.co.kr

**무주반딧불축제**
전북 무주군 무주반딧불축제제전위원회
http://www.firefly.or.kr

**잠자리생태원**
전남 구례군 구례읍 봉서리 891-5 농업기술센터 내
http://www.gurye.jares.go.kr

**국립중앙과학관**
대전광역시 유성구 구성동 32-2
http://www.nsm.go.kr

**한남대학교 자연사박물관**
대전광역시 대덕구 오정동 133
http://museum.hannam.ac.kr

**지당 세계만물박물관**
충남 공주시 탄천면 광명리 104
http://jidang.co.kr

# 찾아보기

## 족(族)명으로 찾기
### Index by Tribe Name

### ㄱ

| | |
|---|---|
| 개미집살이꽃무지족 | 36 |
| 거짓개미집살이꽃무지족 | 36 |
| **골리앗대왕꽃무지족** | **40** |
| 골리앗대왕꽃무지 (비타투스 형) | 50 |
| 골리앗대왕꽃무지 (아피카리스 형) | 46 |
| 골리앗대왕꽃무지 (콘스페르서스 형) | 48 |
| 골리앗대왕꽃무지 | 44 |
| 골리앗대왕흰꽃무지 (푸스투라투스 형) | 51 |
| 골리앗레기우스대왕꽃무지 | 42 |
| 골리앗알보씨그나투스-키르키아누스대왕꽃무지 | 58 |
| 골리앗오리엔탈흰대왕꽃무지 | 52 |
| 골리앗카시쿠스대왕꽃무지 | 60 |
| 골리앗흰대왕꽃무지 (운두라투스 형) | 56 |
| 골리앗흰대왕꽃무지 (프레이시 형) | 54 |
| 구타타기린뿔꽃무지 | 104 |
| 그랄리뿔꽃무지 | 74 |
| 그랄리움부로비타타뿔꽃무지 | 75 |
| 데르비아나오벨투에리왕꽃무지 | 72 |
| 데르비아나왕꽃무지 | 70 |
| 데르비아나콘라드씨왕꽃무지 | 71 |
| 라에타연보석꽃무지 | 111 |
| 레스프렌덴스금광풍뎅이 | 103 |
| 로우예리쌍뿔꽃무지 | 94 |
| 로칠디주걱턱꽃무지 | 112 |
| 롱기셉스앞뿔꽃무지 | 106 |
| 루쑤스왕꽃무지 | 63 |
| 룩케리노랑네점박이앞장다리꽃무지 | 81 |
| 베르토로니흰큰머리꽃무지 | 100 |
| 부르케이셉텐트리오니스앞장다리꽃무지 | 69 |
| 부르케이앞장다리꽃무지 | 68 |
| 사바게이점박이귀신꽃무지 | 83 |
| 사슴풍뎅이 | 66 |
| 셉타노랑띠꽃무지 | 107 |
| 스미씨쉬라티카뿔꽃무지 | 76 |
| 스탄레이넙튠꽃무지 | 96 |
| 스프렌덴스흰큰머리꽃무지 | 101 |
| 아우조욱시뿔꽃무지 | 95 |
| 알페스트리-바후텐시스꽃무지 | 109 |
| 알페스트리스꽃무지 | 108 |
| 오리바세우스사슴뿔꽃무지 | 79 |
| 왈라시사슴풍뎅이 | 65 |
| 유리르히나붉은다리주걱턱꽃무지 | 78 |
| 이리스긴몸광꽃무지 | 110 |
| 주홍대왕귀신꽃무지 (유니칼라 형) | 84 |
| 주홍점박이대왕귀신꽃무지 (데코라타 형) | 85 |
| 줄리아기린뿔꽃무지 | 105 |
| 코로사얼룩뿔꽃무지 | 64 |
| 큐프레오수투랄리스뿔꽃무지 | 73 |
| 크라치지줄무늬귀신꽃무지 | 82 |
| 토르콰타-우간덴시쓰대왕귀신꽃무지 | 88 |
| 토르콰타-이마쿠리콜리스대왕귀신꽃무지 | 86 |
| 토르콰타-포게이대왕귀신꽃무지 | 87 |
| 트릴리네아뿔꽃무지 | 77 |
| 포르나시니왕꽃무지 | 62 |
| 폴리크로우스넙튠꽃무지 | 98 |
| 폴리크로우스넙튠꽃무지 | 97 |
| 풍이 | 102 |
| 하리시-엑스미아큰뿔꽃무지 | 93 |
| 하리시큰뿔꽃무지 | 92 |
| 후라메아-키케리풍이 | 116 |
| 후라메아풍이 | 113 |
| 훼레로이미네티큐앞장다리꽃무지 | 80 |
| **꽃무지족** | **118** |
| 곰꽃무지 | 130 |
| 네점박이붉은띠꽃무지 | 120 |
| 녹스점박이큰꽃무지 | 127 |
| 루마위기점박이꽃무지 | 128 |
| 마르기나타주홍테꽃무지 | 123 |
| 베네라빌리스큰풀색꽃무지 | 137 |
| 샤우미꽃무지 | 138 |
| 스켑시아점박이꽃무지 | 135 |
| 쎄레비카점박이꽃무지 | 126 |
| 에피피아타-환코이시주홍테꽃무지 | 122 |
| 에피피아타-활케이주홍테꽃무지 | 121 |
| 오리엔탈리스점박이꽃무지 | 131 |
| 우후리기큰점박이꽃무지 | 136 |
| 점박이꽃무지 | 133 |
| 쿠프레아-올리바세아꽃무지 | 125 |
| 쿠프레아점박이꽃무지 | 124 |
| 필리핀점박이꽃무지 | 134 |
| 헝가리카꽃무지 | 129 |
| 흰점박이꽃무지 | 132 |

### ㅁ

| | |
|---|---|
| **마다가스카르꽃무지족** | **24** |
| 노랑마다가스카르꽃무지 | 26 |
| **모가슴꽃무지족** | **128** |
| 남미점박이모가슴꽃무지 | 130 |
| 지카니남미점박이모가슴꽃무지 | 131 |

### ㅇ

| | |
|---|---|
| 얼룩꽃무지족 | 144 |
| **오스트레일리아꽃무지족** | **36** |
| 마크래이꽃무지 | 39 |

### ㅌ

| | |
|---|---|
| **투구꽃무지족** | **144** |
| 르히노필루스투구꽃무지 | 146 |
| 볼렌호베니뾰족투구꽃무지 | 149 |
| 스쿠아모수스뾰족투구꽃무지 | 148 |
| 티베타나두뿔투구꽃무지 | 147 |
| 하우드니투구꽃무지 | 150 |

### ㅎ

| | |
|---|---|
| **홀쭉꽃무지족** | **152** |
| 루마위기검은박쥐꽃무지 | 162 |
| 솔로모니카주홍줄홀쭉꽃무지 | 159 |
| 스틱티카홀쭉꽃무지 | 155 |
| 아펠레스주홍줄홀쭉꽃무지 | 158 |
| 앤드로애디홀쭉꽃무지 | 161 |
| 얼룩홀쭉꽃무지 | 154 |
| 엘레강스점박이홀쭉꽃무지 | 156 |
| 트리칼라홀쭉꽃무지 | 163 |
| 프로핀쿠아흰점홀쭉꽃무지 | 160 |
| 후리데리씨금박무늬홀쭉꽃무지 | 157 |

# Index

## 족(族)명으로 찾기
Index by Tribe Name

## C

| | |
|---|---|
| *Cetoniini* | **118** |
| Glycyphana sp. | 120 |
| Pachnoda ephippiata falkei | 121 |
| Pachnoda ephippiata francoisi | 122 |
| Pachnoda marginata | 123 |
| Protaetia brevitarsis seulensis | 132 |
| Protaetia celebica | 126 |
| Protaetia cuprea | 124 |
| Protaetia guam | 130 |
| Protaetia hungarica | 129 |
| Protaetia lumawigi | 128 |
| Protaetia nox | 127 |
| Protaetia orientalis | 131 |
| Protaetia orientalis submarmorea | 133 |
| Protaetia phillppensis | 134 |
| Protaetia scepsia | 135 |
| Protaetia uhligi | 136 |
| Protaetia venerabilis | 137 |
| Proteatia cuprea olivacea | 125 |
| Sternoplus schaumii | 138 |
| *Cremastocheilini* | 36 |

## D

| | |
|---|---|
| *Diplognathini* | 144 |

## G

| | |
|---|---|
| *Goliathini* | **40** |
| Cheirolasia burkei burkei | 68 |
| Cheirolasia burkei septentrionis | 69 |
| Cyphonocephalus olivaceus | 79 |
| Dicarnocephalus adamsi | 66 |
| Dicranocephalus wallichii | 65 |
| Dicronorrhina derbyana conrads | 71 |
| Dicronorrhina derbyana derbyana | 70 |
| Dicronorrhina derbyana oberthuer | 72 |
| Eudicella cupreosuturalis | 73 |
| Eudicella gralli gralli | 74 |
| Eudicella gralli umbrovittata | 75 |
| Eudicella smithi shiratica | 76 |
| Eudicella trilineata | 77 |
| Fornasinius fornasinii | 62 |
| Fornasinius russus | 63 |
| Goliathus albosignatus kirkianus | 50 |
| Goliathus cacicus | 58 |
| Goliathus Goliathus | 60 |
| Goliathus goliathus - Form apicalis | 44 |
| Goliathus goliathus - Form conspersus | 46 |
| Goliathus goliathus - Form vittatus | 48 |
| Goliathus orientalis - Form preissi | 54 |
| Goliathus orientalis - Form pustulatus | 51 |
| Goliathus orientalis - Form undulatus | 56 |
| Goliathus orientalis | 52 |
| Goliathus regius | 42 |
| Hypselogenia corrosa | 64 |
| Ingrisma euryrrhina | 78 |
| Jumnos ferreroiminettiique | 80 |
| Jumnos ruckeri | 81 |
| Mecynorhina kraatzi | 82 |
| Mecynorhina oberthueri - Form decorata | 85 |
| Mecynorhina oberthueri -Form unicolor | 84 |
| Mecynorhina savagei | 83 |
| Mecynorhina torquata immaculicollis | 86 |
| Mecynorhina torquata poggei | 87 |
| Mecynorhina torquata ugandensis | 88 |
| Megalorhina harrisi eximia | 93 |
| Megalorhina harrisi harrisihi | 92 |
| Mystroceros rouyeri | 94 |
| Neophaedimus auzouxi | 95 |
| Neptunides polychrous | 98 |
| Neptunides polychrous polychrous | 97 |
| Neptunides stanleyi | 96 |
| Pedinorrhina septa | 107 |
| Pseudotorynorrhina japonica | 102 |
| Rhamphorrhina bertolonii | 100 |
| Rhamphorrhina splendens | 101 |
| Rhomborhina resplendens | 103 |
| Stephanorrhina julia | 105 |
| Stephanorrhina guttata | 104 |
| Taurrhina longiceps | 106 |
| Tmesorrhina alpestris | 108 |
| Tmesorrhina alpestris bafutensis | 109 |
| Tmesorrhina iris | 110 |
| Tmesorrhina laeta | 111 |
| Torynorrhina flammea chicheryi | 116 |
| Torynorrhina flammea flammea | 113 |
| Trigonophorus rothschildi | 112 |
| *Gymnetini* | **140** |
| Gymnetis pantherina | 142 |
| Gymnetis pantherina zikani | 143 |

## P

| | |
|---|---|
| *Phaedimini* | **144** |
| Mycteristes rhinophyllus | 146 |
| Mycteristes squamosus | 148 |
| Mycteristes tibetana | 147 |
| Mycteristes vollenhoveni | 149 |
| Phaedimus howdeni | 150 |

## S

| | |
|---|---|
| *Schizorhinini* | **36** |
| Trichaulax macleayi | 39 |
| *Stenotarsiini* | **36** |
| Euchroea auripimenta | 38 |

## T

| | |
|---|---|
| *Taenioderini* | **152** |
| Euselates sp. | 154 |
| Euselates stictica | 155 |
| Ixorida elegans | 156 |
| Ixorida friderici | 157 |
| Ixorida propinqua | 160 |
| Ixorida solomonica | 159 |
| Ixorida venerea apelles | 158 |
| Plectrone endroedii | 161 |
| Plectrone lumawigi | 162 |
| Taeniodera tricolor tricolor | 163 |

## X

| | |
|---|---|
| *Xiphoscelidini* | 36 |

# 찾아보기

## 속명으로 찾기
### Index by Generic name

### ㄱ

| | |
|---|---|
| 골리앗대왕꽃무지 (비타투스 형) | 50 |
| 골리앗대왕꽃무지 (아피카리스 형) | 46 |
| 골리앗대왕꽃무지 (콘스페르수스 형) | 48 |
| 골리앗대왕꽃무지 | 44 |
| 골리앗대왕흰꽃무지 (푸스투라투스 형) | 51 |
| 골리앗레기우스대왕꽃무지 | 42 |
| 골리앗알보씨그나투스-키르키아누스대왕꽃무지 | 58 |
| 골리앗오리엔탈흰대왕꽃무지 | 52 |
| 골리앗카시쿠스대왕꽃무지 | 60 |
| 골리앗흰대왕꽃무지 (운두라투스 형) | 56 |
| 골리앗흰대왕꽃무지 (프레이시 형) | 54 |
| 괌꽃무지 | 130 |
| 구타타기린뿔꽃무지 | 104 |
| 그랄리뿔꽃무지 | 74 |
| 그랄리움부로비타타뿔꽃무지 | 75 |

### ㄴ

| | |
|---|---|
| 남미점박이모가슴꽃무지 | 142 |
| 네점박이붉은띠꽃무지 | 120 |
| 노랑마다가스카르꽃무지 | 38 |
| 녹스점박이큰꽃무지 | 127 |

### ㄷ

| | |
|---|---|
| 데르비아나오벨투에리왕꽃무지 | 72 |
| 데르비아나왕꽃무지 | 70 |
| 데르비아나콘라드씨왕꽃무지 | 71 |

### ㄹ

| | |
|---|---|
| 라에타연보석꽃무지 | 111 |
| 레스프렌덴스금광풍뎅이 | 103 |
| 로우예리쌍뿔꽃무지 | 94 |
| 로칠드주걱턱꽃무지 | 112 |
| 롱기셉스앞뿔꽃무지 | 106 |
| 루마위기검은박쥐꽃무지 | 162 |
| 루마위기점박이꽃무지 | 128 |
| 루쑤스왕꽃무지 | 63 |
| 룩케리노랑네점박이앞장다리꽃무지 | 81 |
| 르히노필루스투구꽃무지 | 146 |

### ㅁ

| | |
|---|---|
| 마르기나타주홍꽃무지 | 123 |
| 마크래이꽃무지 | 39 |

### ㅂ

| | |
|---|---|
| 베네라빌리스큰풀색꽃무지 | 137 |
| 베르토니흰큰머리꽃무지 | 100 |
| 볼렌호베니뾰족투구꽃무지 | 149 |
| 부르케이셉텐트리오니스앞장다리꽃무지 | 69 |
| 부르케이앞장다리꽃무지 | 68 |

### ㅅ

| | |
|---|---|
| 사바게이점박이귀신꽃무지 | 83 |
| 사슴풍뎅이 | 66 |
| 샤우미꽃무지 | 138 |
| 셉타노랑띠꽃무지 | 107 |
| 솔로모니카주홍줄홀쭉꽃무지 | 159 |
| 스미씨쉬라티카뿔꽃무지 | 76 |
| 스켑시아점박이꽃무지 | 135 |
| 스쿠아모수스뾰족투구꽃무지 | 148 |
| 스탄레이넙튠꽃무지 | 96 |
| 스틱티카홀쭉꽃무지 | 155 |
| 스프레덴스흰큰머리꽃무지 | 101 |
| 쎄레비카점박이꽃무지 | 126 |

### ㅇ

| | |
|---|---|
| 아우조욱시뿔꽃무지 | 95 |
| 아펠레스주홍줄홀쭉꽃무지 | 158 |
| 알페스트리-바후텐시스꽃무지 | 109 |
| 알페스트리스꽃무지 | 108 |
| 앤드로애디홀쭉꽃무지 | 161 |
| 얼룩홀쭉꽃무지 | 154 |
| 에피피아타-환코이시주홍테꽃무지 | 122 |
| 에피피아타-활케이주홍테꽃무지 | 121 |
| 엘레강스점박이홀쭉꽃무지 | 156 |
| 오리바세우스사슴꽃무지 | 79 |
| 오리엔탈리스점박이꽃무지 | 131 |
| 왈라치사슴풍뎅이 | 65 |
| 우후리기큰점박이꽃무지 | 136 |
| 유리르히나붉은다리주걱턱꽃무지 | 78 |
| 이리스긴몸광꽃무지 | 110 |

### ㅈ

| | |
|---|---|
| 점박이꽃무지 | 133 |
| 주홍대왕귀신꽃무지 (유니칼라 형) | 84 |
| 주홍점박이대왕귀신꽃무지 (데코라타 형) | 85 |
| 줄리아기린뿔꽃무지 | 105 |
| 지카니남미점박이모가슴꽃무지 | 143 |

### ㅋ

| | |
|---|---|
| 코로사얼룩뿔꽃무지 | 64 |
| 쿠프레아-올리바세아꽃무지 | 125 |
| 쿠프레아점박이꽃무지 | 124 |
| 큐프레오수투랄리스뿔꽃무지 | 73 |
| 크라치지줄무늬귀신꽃무지 | 82 |

### ㅌ

| | |
|---|---|
| 토르코타-우간덴시쓰대왕귀신꽃무지 | 88 |
| 토르코타-이마쿠리콜리스대왕귀신꽃무지 | 86 |
| 토르코타-포게이대왕귀신꽃무지 | 87 |
| 트리칼라홀쭉꽃무지 | 163 |
| 트릴리네아뿔꽃무지 | 77 |
| 티베타나두뿔투구꽃무지 | 147 |

### ㅍ

| | |
|---|---|
| 포르나시니왕꽃무지 | 62 |
| 폴리크로우스넙튠꽃무지 | 98 |
| 폴리크로우스넙튠꽃무지 | 97 |
| 풍이 | 102 |
| 프로핀쿠아흰점홀쭉꽃무지 | 160 |
| 필리핀점박이꽃무지 | 134 |

### ㅎ

| | |
|---|---|
| 하리시-엑스미아큰뿔꽃무지 | 93 |
| 하리시큰뿔꽃무지 | 92 |
| 하우드니투구꽃무지 | 150 |
| 헝가리카꽃무지 | 129 |
| 후라메아-키케리풍이 | 116 |
| 후라메아풍이 | 113 |
| 후리데리씨금박무늬홀쭉꽃무지 | 157 |
| 훼레로이미네티큐앞장다리꽃무지 | 80 |
| 흰점박이꽃무지 | 132 |

# Index

## 속명으로 찾기
### Index by Generic name

### C

| | |
|---|---|
| *Cheirolasia burkei burkei* | 68 |
| *Cheirolasia burkei septentrionis* | 69 |
| *Cyphonocephalus olivaceus* | 79 |

### D

| | |
|---|---|
| *Dicarnocephalus adamsi* | 66 |
| *Dicranocephalus wallichii* | 65 |
| *Dicronorrhina derbyana conrads* | 71 |
| *Dicronorrhina derbyana derbyana* | 70 |
| *Dicronorrhina derbyana oberthuer* | 72 |

### E

| | |
|---|---|
| *Euchroea auripimenta* | 38 |
| *Eudicella cupreosuturalis* | 73 |
| *Eudicella gralli gralli* | 74 |
| *Eudicella gralli umbrovittata* | 75 |
| *Eudicella smithi shiratica* | 76 |
| *Eudicella trilineata* | 77 |
| *Euselates sp.* | 154 |
| *Euselates stictica* | 155 |

### F

| | |
|---|---|
| *Fornasinius fornasinii* | 62 |
| *Fornasinius russus* | 63 |

### G

| | |
|---|---|
| *Glycyphana sp.* | 120 |
| *Goliathus goliathus - Form vittatus* | 50 |
| *Goliathus albosignatus kirkianus* | 58 |
| *Goliathus cacicus* | 60 |
| *Goliathus goliathus* | 44 |
| *Goliathus goliathus - Form apicalis* | 46 |
| *Goliathus goliathus - Form conspersus* | 48 |
| *Goliathus orientalis - Form preissi* | 54 |
| *Goliathus orientalis - Form pustulatus* | 51 |
| *Goliathus orientalis - Form undulatus* | 56 |
| *Goliathus orientalis* | 52 |
| *Goliathus regius* | 42 |
| *Gymnetis pantherina* | 142 |
| *Gymnetis pantherina zikani* | 143 |

### H

| | |
|---|---|
| *Hypselogenia corrosa* | 64 |

### I

| | |
|---|---|
| *Ingrisma euryrrhina* | 78 |
| *Ixorida elegans* | 156 |
| *Ixorida friderici* | 157 |
| *Ixorida propinqua* | 160 |
| *Ixorida solomonica* | 159 |
| *Ixorida venerea apelles* | 158 |

### J

| | |
|---|---|
| *Jumnos ferreroiminettiique* | 80 |
| *Jumnos ruckeri* | 81 |

### M

| | |
|---|---|
| *Mecynorhina kraatzi* | 82 |
| *Mecynorhina oberthueri - Form decorata* | 85 |
| *Mecynorhina oberthueri - Form unicolor* | 84 |
| *Mecynorhina savagei* | 83 |
| *Mecynorhina torquata immaculicollis* | 86 |
| *Mecynorhina torquata poggei* | 87 |
| *Mecynorhina torquata ugandensis* | 88 |
| *Megalorhina harrisi eximia* | 93 |
| *Megalorhina harrisi harrisihi* | 92 |
| *Mycteristes rhinophyllus* | 146 |
| *Mycteristes squamosus* | 148 |
| *Mycteristes tibetana* | 147 |
| *Mycteristes vollenhoveni* | 149 |
| *Mystroceros rouyeri* | 94 |

### N

| | |
|---|---|
| *Neophaedimus auzouxi* | 95 |
| *Neptunides polychrous* | 98 |
| *Neptunides polychrous polychrous* | 97 |
| *Neptunides stanleyi* | 96 |

### P

| | |
|---|---|
| *Pachnoda ephippiata falkei* | 121 |
| *Pachnoda ephippiata francoisi* | 122 |
| *Pachnoda marginata* | 123 |
| *Pedinorrhina septa* | 107 |
| *Phaedimus howdeni* | 150 |
| *Plectrone endroedii* | 161 |
| *Plectrone lumawigi* | 162 |
| *Protaetia brevitarsis seulensis* | 132 |
| *Protaetia celebica* | 126 |
| *Protaetia cuprea* | 124 |
| *Protaetia guam* | 130 |
| *Protaetia hungarica* | 129 |
| *Protaetia lumawigi* | 128 |
| *Protaetia nox* | 127 |
| *Protaetia orientalis* | 131 |
| *Protaetia orientalis submarmorea* | 133 |
| *Protaetia phillppensis* | 134 |
| *Protaetia scepsia* | 135 |
| *Protaetia uhligi* | 136 |
| *Protaetia venerabilis* | 137 |
| *Proteatia cuprea olivacea* | 125 |
| *Pseudotorynorrhina japonica* | 104 |

### R

| | |
|---|---|
| *Rhamphorrhina bertolonii* | 100 |
| *Rhamphorrhina splendens* | 101 |
| *Rhomborhina resplendens* | 103 |

### S

| | |
|---|---|
| *Stephanorrhina julia* | 105 |
| *Stephanorrhina guttata* | 104 |
| *Sternoplus schaumii* | 138 |

### T

| | |
|---|---|
| *Taeniodera tricolor tricolor* | 163 |
| *Taurrhina longiceps* | 106 |
| *Tmesorrhina alpestris* | 108 |
| *Tmesorrhina alpestris bafutensis* | 109 |
| *Tmesorrhina iris* | 110 |
| *Tmesorrhina laeta* | 111 |
| *Torynorrhina flammea chicheryi* | 116 |
| *Torynorrhina flammea flammea* | 113 |
| *Trichaulax macleayi* | 39 |
| *Trigonophorus rothschildi* | 112 |

# 찾아보기

## 종명으로 찾기
Index by Species name

### A

| | |
|---|---|
| adamsi, Dicarnocephalus | 66 |
| albosignatus kirkianus, Goliathus | 58 |
| alpestris, Tmesorrhina | 109 |
| alpestris bafutensis, Tmesorrhina | 108 |
| auripimenta, Euchroea | 38 |
| auzouxi, Neophaedimus | 95 |

### B

| | |
|---|---|
| bertolonii, Rhamphorrhina | 100 |
| brevitarsis seulensis, Protaetia | 132 |
| burkei burkei, Cheirolasia | 68 |
| burkei septentrionis, Cheirolasia | 69 |

### C

| | |
|---|---|
| cacicus, Goliathus | 60 |
| celebica, Protaetia | 126 |
| corrosa, Hypselogenia | 64 |
| cuprea, Protaetia | 125 |
| cuprea olivacea, Proteatia | 124 |
| cupreosuturalis, Eudicella | 73 |

### D

| | |
|---|---|
| derbyana conradsi, Dicronorrhina | 71 |
| derbyana derbyana, Dicronorrhina | 70 |
| derbyana oberthueri, Dicronorrhina | 72 |

### E

| | |
|---|---|
| elegans, Ixorida | 156 |
| endroedii, Plectrone | 161 |
| ephippiata falkei, Pachnoda | 121 |
| ephippiata francoisi, Pachnoda | 122 |
| euryrrhina, Ingrisma | 78 |

### F

| | |
|---|---|
| ferreroiminettiique, Jumnos | 80 |
| flammea chicheryi, Torynorrhina | 116 |
| flammea flammea, Torynorrhina | 115 |
| fornasinii, Fornasinius | 62 |
| friderici, Ixorida | 157 |

### G

| | |
|---|---|
| goliathus, Goliathus | 46 |
| goliathus - Form apicalis, Goliathus | 48 |
| goliathus - Form conspersus, Goliathus | 50 |
| goliathus - Form vittatus, Goliathus | 44 |
| gralli gralli, Eudicella | 74 |
| gralli umbrovittata, Eudicella | 75 |
| guam, Protaetia | 130 |
| guttata, Stephanorrhina | 104 |

### H

| | |
|---|---|
| harrisi eximia, Megalorhina | 93 |
| harrisi harrisihi, Megalorhina | 92 |
| howdeni, Phaedimus | 150 |
| hungarica, Protaetia | 129 |

### I

| | |
|---|---|
| iris, Tmesorrhina | 110 |

### J

| | |
|---|---|
| japonica, Pseudotorynorrhina | 90 |
| julia, Stephanorrhina | 93 |

### K

| | |
|---|---|
| kraatzi, Mecynorhina | 82 |

### L

| | |
|---|---|
| laeta, Tmesorrhina | 111 |
| longiceps, Taurrhina | 106 |
| lumawigi, Plectrone | 162 |
| lumawigi, Protaetia | 128 |

### M

| | |
|---|---|
| macleayi, Trichaulax | 38 |
| marginata, Pachnoda | 123 |

### N

| | |
|---|---|
| nox, Protaetia | 127 |

### O

| | |
|---|---|
| oberthueri - Form decorata, Mecynorhina | 85 |
| oberthueri - Form unicolor, Mecynorhina | 84 |
| olivaceus, Cyphonocephalus | 79 |
| orientalis, Protaetia | 54 |
| orientalis - Form preissi, Goliathus | 51 |
| orientalis - Form pustulatus, Goliathus | 56 |
| orientalis - Form undulatus, Goliathus | 133 |
| orientalis, Goliathus | 52 |
| orientalis submarmorea, Protaetia | 131 |

### P

| | |
|---|---|
| pantherina, Gymnetis | 143 |
| pantherina zikani, Gymnetis | 142 |
| phillppensis, Protaetia | 134 |
| polychrous, Neptunides | 97 |
| polychrous polychrous, Neptunides | 98 |
| propinqua, Ixorida | 160 |

### R

| | |
|---|---|
| regius, Goliathus | 42 |
| resplendens, Rhomborhina | 103 |
| rhinophyllus, Mycteristes | 146 |
| rothschildi, Trigonophorus | 112 |
| rouyeri, Mystroceros | 94 |
| ruckeri, Jumnos | 81 |
| russus, Fornasinius | 63 |

### S

| | |
|---|---|
| savagei, Mecynorhina | 83 |
| scepsia, Protaetia | 135 |
| schaumii, Sternoplus | 138 |
| septa, Pedinorrhina | 107 |
| smithi shiratica, Eudicella | 76 |
| solomonica, Ixorida | 159 |
| sp., Glycyphana | 154 |
| sp., Euselates | 120 |
| splendens, Rhamphorrhina | 101 |
| squamosus, Mycteristes | 148 |
| stanleyi, Neptunides | 96 |
| stictica, Euselates | 155 |

# Index

## T

*tibetana, Mycteristes* — 147
*torquata immaculicollis, Mecynorhina* — 86
*torquata poggei, Mecynorhina* — 87
*torquata ugandensis, Mecynorhina* — 88
*tricolor tricolor, Taeniodera* — 163
*trilineata, Eudicella* — 77

## U

*uhligi, Protaetia* — 136

## V

*venerabilis, Protaetia* — 137
*venerea apelles, Ixorida* — 158
*vollenhoveni, Mycteristes* — 149

## W

*wallichii, Dicranocephalus* — 65

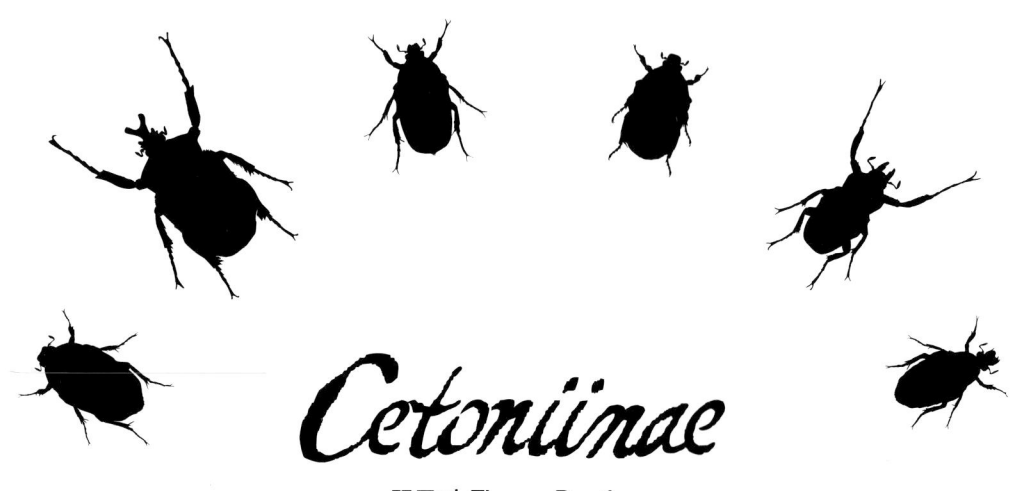

# Cetoniinae
꽃무지 | Flower Beetles

**세계유용곤충대도감시리즈**
Photograph Book Series of the World Insects

Vol. Ⅰ 세계의 사슴벌레 대도감  The Lucanid Beetles of the World (Stag Beetles)
Vol. Ⅱ 세계의 장수풍뎅이 대도감  The Dynastid Beetles of the World (Rhinoceros Beetles)
**Vol. Ⅲ 세계의 꽃무지 대도감  The Cetoniid Beetles of the World (Flower Beetles)**
Vol. Ⅳ 세계의 하늘소 대도감  The Cerambycid Beetles of the World (Longicorn Beetles)
Vol. Ⅴ 세계의 진귀곤충 대도감  The Rare Beetles of the World
Vol. Ⅵ 전북의 곤충 대도감  The Insects of Jeonbuk in Korea
Vol. Ⅶ 부안의 곤충 대도감  The Insects of Buan in Korea